MILITARY MENTAL TOUGHNESS

MILITARY MENTAL TOUGHNESS

ELITE TRAINING FOR CRITICAL SITUATIONS

CHRIS MCNAB

Ulysses Press

Publishers in the United States by:
Ulysses Press
P.O. Box 3440
Berkeley, CA 94703
www.ulyssespress.com

First published in 2016 in the United Kingdom as *Mental Fitness* by Amber Books Ltd.

ISBN: 978-1-61243-603-6
Library of Congress Control Number: 2016934495

Project Editor: Michael Spilling
U.S. Editor: Renee Rutledge
Designer: Brian Rust
Illustrator: Tony Randell
All illustrations © Amber Books Ltd.
Cover designer: what!design @ whatweb.com
Cover photograph: © aaltair/shutterstock.com

Distributed by Publishers Group West

Printed in China

NOTE TO READERS
This book has been written and published strictly for informational purposes. The author and publisher are providing you with information in this work so that you can have the knowledge and can choose, at your own risk, to act on that knowledge. Neither the author or publisher can accept responsibility for any loss, injury, or damage caused as a result of the use of techniques described in this book, nor for any prosecutions or proceedings brought or instigated against any person or body that may result from using these techniques. This book is sold without warranties or guarantees of any kind, and the author and publisher disclaim any responsibility or liability for personal injury, property damage, or any other loss or damage, however caused, related to the information in this book.

CONTENTS

INTRODUCTION

Over the last half-century, there has been a vast amount of research conducted into what separates an effective military force—one that largely achieves its objectives efficiently and with minimal casualties—from a poor-performing one. Of course, there are multiple factors complicating this equation. Financial wealth (in terms of defense spending), investment in technology, national morale, living conditions of military personnel, sheer manpower, available firepower—all these issues and more bear upon a military unit's capability to do a job and survive in the process.

With unsophisticated reasoning, it might appear that the strongest army simply has the most manpower, the largest arsenal and the greatest financial backing. While it would be churlish to deny that these factors are not important in the total power projection of an armed force, sheer scale is not by any means the final arbiter of victory. Rather, what often seems to separate victorious from defeated armies is, arguably, *intelligence*. By "intelligence" I do not mean the professional collection and processing of tactical and strategic information by specialist units, who gather HUMINT (human intelligence) and SIGINT (signals intelligence). Instead, I refer to the

A military unit conducts an urban patrol. Mutual trust in each other's abilities is essential for the development of unit cohesion.

ability of all members of a military unit or formation, from the most senior commander down to the lowliest private, to manage their mental states, design rational and informed methods of achieving their objectives, implement them rapidly and with aggression, and deal resourcefully with the multiple problems that afflict them in both peacetime and wartime.

In 1997, the U.S. Marine Corps issued an updated version of its seminal field manual publication *Warfighting*, a short and insightful book into the theory and practice of war at its highest level. The book is full of truly profound insights, not

only for serving military personnel, but also for all those interested in human performance in extreme circumstances. In one enlightening passage, the author spells out the relationship between human character and the outcome of conflict:

> *War is an extreme trial of moral and physical strength and stamina. Any view of the nature of war would hardly be accurate or complete without consideration of the effects of danger, fear, exhaustion and*

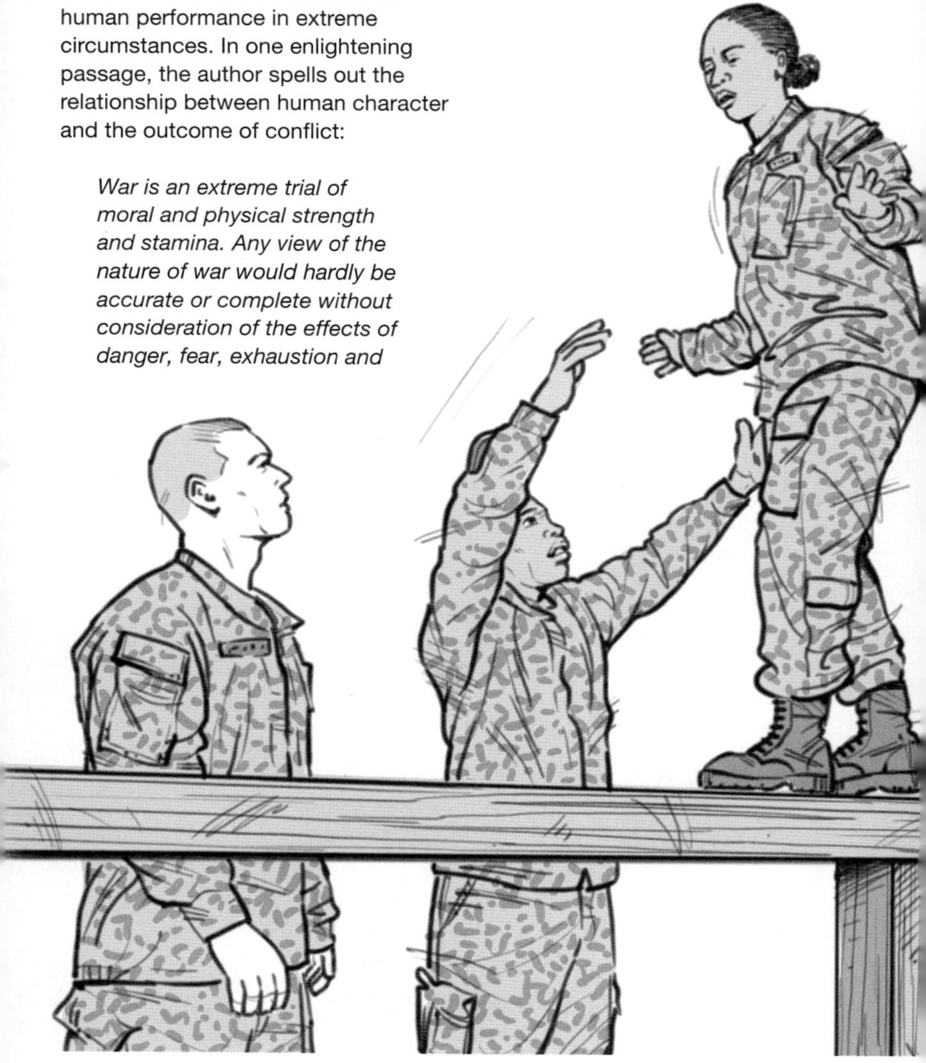

privation on those who must do the fighting. However, these effects vary greatly from case to case. Individuals and peoples react differently to the stress of war; an act that may break the will of one enemy may only serve to stiffen the resolve of another. Human will, instilled through leadership, is the driving force of all action in war. No degree of technological development or scientific calculation will diminish the human dimension in war. Any doctrine which attempts to reduce warfare to ratios of forces, weapons and equipment neglects the impact of the human will on the conduct of war and is therefore inherently flawed.

—U.S. Marine Corps, MCDP 1, *Warfighting,* pp.12–13

This book is, in essence, a reflection on "the impact of human will" in military circles, also taking the lessons learned and applying them to the broader experience of life. For the military has much to teach those who do not belong to it. Its men and women must confront situations that stretch human character to its very limits, from absolute and prolonged boredom to sudden and shocking terror. Because of these demands placed on the character of its personnel, professional armed services have invested heavily in the study of human psychology, divulging many of its findings in doctrinal publications, which we will reference throughout the book. Here we will explore the military mindset and discover how to change our own thinking in the process.

Military training is not just about individual physical and mental development, but also about learning to trust a team in the decision-making process.

Oil
Milk

1

Mind and body

The level of interaction between mind and body is profound. Your physical state, to a large degree, dictates how you think, feel, and behave.

In 2015, reports began to appear in the international media of a pervasive new challenge facing military recruiters in the United States. It was revealed that almost one in three people aged between 17 and 24 was ineligible for military service on account of obesity. Indeed, obesity was one of the leading reasons why military applications were rejected. (For reference, poor education, a criminal record and drug use were the other principal factors.)

Obesity, as we now know, is a profound medical problem facing modern societies around the world. We live in what has been termed an "obesogenic environment," surroundings that promote the intake of excessive calories while also preventing, or at least retarding, efforts to lose weight. A snapshot view of a typical main street or shopping mall reveals the truth behind this viewpoint. Food temptation assails us almost every few feet, from restaurants, cafes, fast-food outlets, and supermarkets—many of them using proven psychological techniques to tap into our hard-wired evolutionary instinct to eat plenty while plenty is available. An adult who succumbs to these temptations (and most of us

Opposite: Our food, environment and sleep patterns all have an influence over our mental lives.

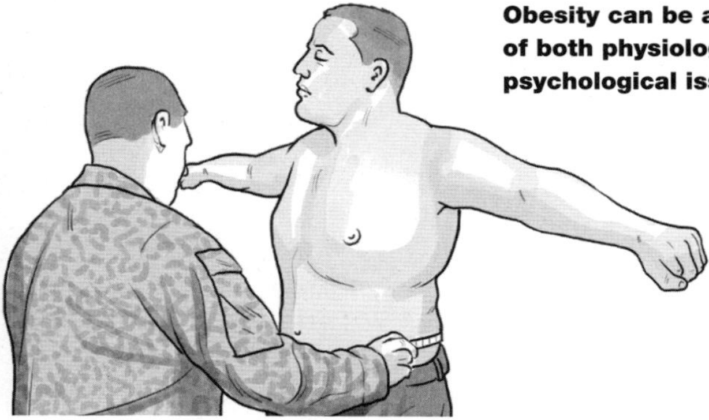

Obesity can be a warning sign of both physiological and also psychological issues.

do) can exceed the recommended daily calorie intake within a matter of minutes by eating foods heavy in fat, sugar and salt, piling on the pounds. This problem has even infiltrated the military world. Many armed services have seen a steady rise in the numbers of military personnel discharged from the service on account of their obesity. The fact remains that to serve in the armed forces, you have to be fit to fight.

Body First

It might seem unusual to begin with a reflection upon obesity, but actually it connects with one of the most critical lessons that we will make throughout this book: *What we perceive as spiritual or psychological problems might in many cases simply be physical or practical problems that need to be resolved*. Obesity is a case in point. Just on a physical level, obesity reduces our energy levels and the efficient functioning of core processes, such as heartbeat, digestion and breathing. Little wonder, therefore, that depression can be commonplace among obese people; not only are the obese coping with the regular trials and tribulations of living, but their bodies have been geared to deplete their resilience and stamina. Furthermore, obesity can

Eating meals together has definite social benefits, but don't lose track of your food consumption.

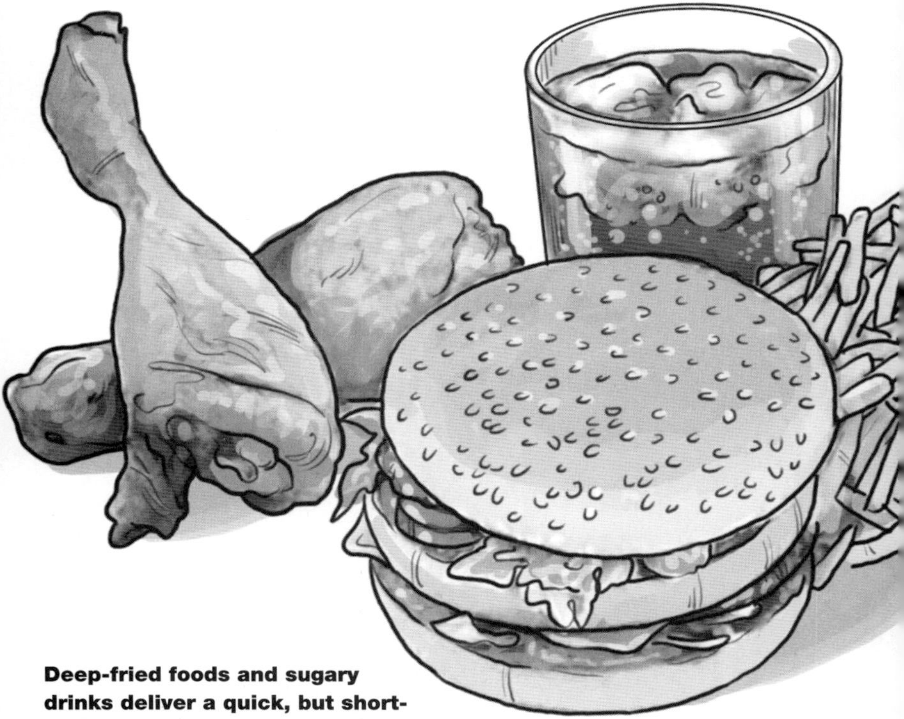

Deep-fried foods and sugary drinks deliver a quick, but short-lasting, energy boost. Consume them only in moderation.

prevent them from doing some of the very activities, such as physical exercise, that would enable them to cope better with side effects, such as depression. On top of all that come the general social problems of obesity, fueled by a loss of self-esteem.

Not all of the readers of this book will suffer from obesity problems; indeed, many may be extremely fit and have excellent nutrition. Yet the principle remains valid even for those readers—the mind and the body are inextricably linked, and one of the most valuable investments you can make for your mental health is to improve your physical health.

This is where the armed services excel. Every year they take hundreds of thousands of people, from all walks of life and with all standards of fitness, and transform them into individuals of near-athlete standards of fitness. So, the first step in getting into the military mindset is to get physically fit. What precisely do we mean by physical fitness? The U.S. Army's *Physical Fitness Training* manual clarifies:

Military physical training (PT) invests heavily in muscular development. This tire-flipping exercise promotes whole-body strength while encouraging a determined attitude.

The essential food groups (clockwise from top right): fats, carbohydrates, fruits and vegetables, sources of vitamins and minerals, and proteins.

The components of physical fitness are as follows:

- *Cardiorespiratory (CR) endurance—the efficiency with which the body delivers oxygen and nutrients needed for muscular activity and transports waste products from the cells.*
- *Muscular strength—the greatest amount of force a muscle or muscle group can exert in a single effort.*
- *Muscular endurance—the ability of a muscle or muscle group to*

Professional military units set up strict schedules of hydration in the field. Some of the early signs of dehydration are negative mood swings.

perform repeated movements with a sub-maximal force for extended periods of time.

- *Flexibility—the ability to move the joints (for example, elbow, knee) or any group of joints through an entire, normal range of motion.*
- *Body composition—the amount of body fat a soldier has in comparison to his total body mass.*

—FM 21-20, *Physical Fitness Training,* 1-3

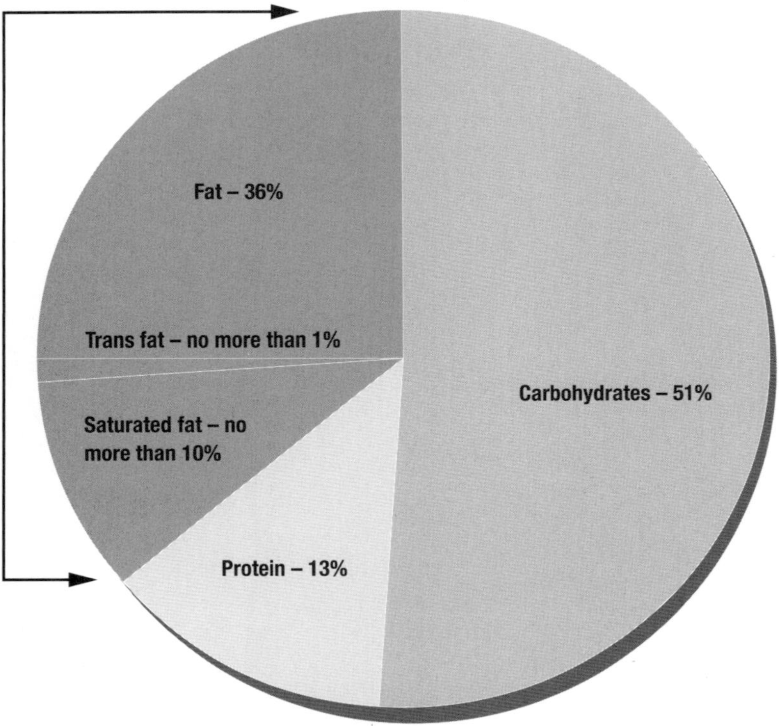

This pie chart shows a representation of a recommended balanced diet in terms of protein, carbohydrate and fat intake.

To this list we should add the requirement for a healthy diet. By this we mean a diet that includes all essential vitamins, minerals, carbohydrates, fats and sugars, and in properly balanced and moderated amounts. Such a diet is heavily biased toward fresh fruit and vegetables, lean meats (with a reduction in red meat intake), low intake of junk food and candy, and a calorie intake appropriate to your level of physical exertion. Thus an adult man requires on average 2500 calories a day as a recommended intake, but this can climb much higher during periods of intense daily physical activity. During Royal Marines training, for example, the new soldiers can be consuming as many as 4000 calories a day and yet still lose weight, such is the pressure they are putting their bodies under.

This book is not the place to explain in detail a physical exercise program, nor the intricacies of a good-quality diet. We can, however, make some general and invaluable points about physical conditioning, based upon military development programs:

Develop a Fitness Program

Rather than think, "I must get fit," instead, set clear, progressive and ultimately achievable fitness targets, and pursue them with single-minded purpose on a weekly basis. As occurs within the armed forces, you must build fitness into your lifestyle rather than think of it as an adjunct to your day. Furthermore, in the military every individual, including those in non-physical administrative positions, has to maintain the core levels of fitness and undergo regular physical conditioning. Likewise, even if your work does not entail heavy physical activity, consider your fitness program to be an essential part of maintaining your peak performance, both mental and physical.

It is also critical that you set yourself definable goals within your fitness regimen. The physical fitness entry requirements for most military units are easily found online, and you can use these as goals for your physical training. Military targets are particularly useful because they focus on a well-rounded physical development, promoting

Some sweets can be eaten for an instant energy boost, but they should be eaten infrequently and never as a replacement for meals or proper nutrition.

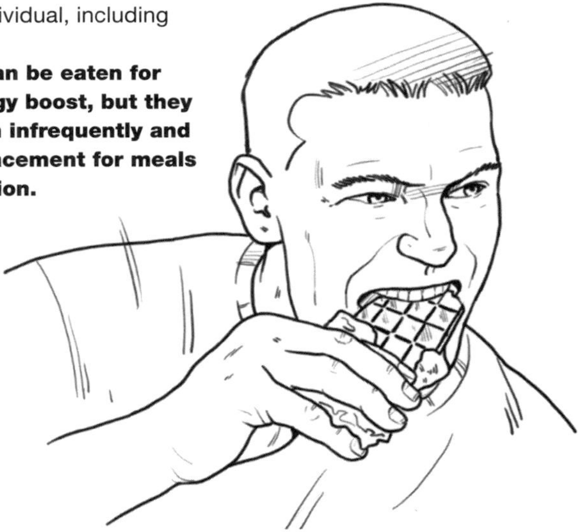

Ranger Physical Fitness Test (PFT)	Minimum Scores	Recommended Scores
Push-ups in 2:00	49 in 2:00	80
Sit-ups in 2:00	59	80
Pull-ups	6	12
2-mile [3.2-km] run	15:12	Sub 13:00
5-mile [8-km] run	40:00	35:00
16-mile [25.7-km] hike w/65lb [29.5kg] pack	5 hours 20 minutes	4–5 hours
15-meter [49-ft] swim with gear	Pass/Fail	Pass/Fail

people suitable for Ranger training in the United States.

As we can see here, there are mixed tests involved, ranging from simple strength exercises, such as push-ups, to swimming while weighed down with gear. Similarly, you should build a mixed variety of activities into your fitness regimen. A good balance would be one or two runs a week (cardiovascular), one session of circuit training (strength and cardiovascular),

Soldiers' communal eating habits have numerous psychological benefits, such as stress relief, relaxation and information sharing.

and one session of weight training (strength). An alternative could be one run and one swim each week, plus two sessions in the gym. Yoga can also be a valuable addition for promoting flexibility, a quality that tends to be undervalued by many people preparing for military training. Interestingly, one problem experienced by many Western soldiers operating in Afghanistan has been that they find it difficult to sit for long periods cross-legged on the floor, talking with local people who are entirely comfortable in such a position. So a simple lack of flexibility can make it harder to conduct the "hearts and minds" operations so essential in that part of the world.

Enjoy Your Fitness

You are unlikely to maintain a fitness regimen if you do not fundamentally enjoy the types of activity you have chosen to do. There are numerous different sports out there, so make sure you experiment to find the ones that really give you a "buzz," and that either promote a restorative peace of mind (typically solitary activities such as country runs or rowing) or the social benefits of team sports. Furthermore, most activities lead to the release of endorphins, pleasure-producing chemicals in the body, thus regular physical activity has been clinically proven to reduce both stress and depression, even in the most severely affected patients.

Under operational conditions, sleep might be limited. Twenty-minute naps every four hours can sustain energy in the short term.

Watch Out for Obsession

Sometimes people can form a rather addictive relationship with physical exercise in which they feel they have to do intense physical activity every day. Otherwise, they experience feelings of guilt and self-punishment. While military units can and do put soldiers under extreme and prolonged physical pressures, mainly during training exercises and combat operations, such regimens are never maintained indefinitely, and will be punctuated with periods of rest and recuperation. The Royal Marines provide advice for anybody wanting to get fit to apply for service. They suggest that if you do two training sessions per day you must include a rest session in between of several hours' duration, and have a meal to replenish energy levels. Most important, however, is that they state you must have at least one day a week in which you do no physical training whatsoever. This rest period is critical to allow your body to strengthen and to recover from the exercise; constant overexercising will lead to injury, high levels of fatigue and a weakened immune system. Furthermore, if exercise builds into an uncontrollable compulsion, it can actually lead to depressive symptoms—the very thing that exercise is meant to cure.

Support with Proper Nutrition

Finally, make sure that you do not undo the benefits of your exercise regimen by eating poorly. The U.S.

Sleep deprivation produces a broad range of detrimental physical and psychological effects.

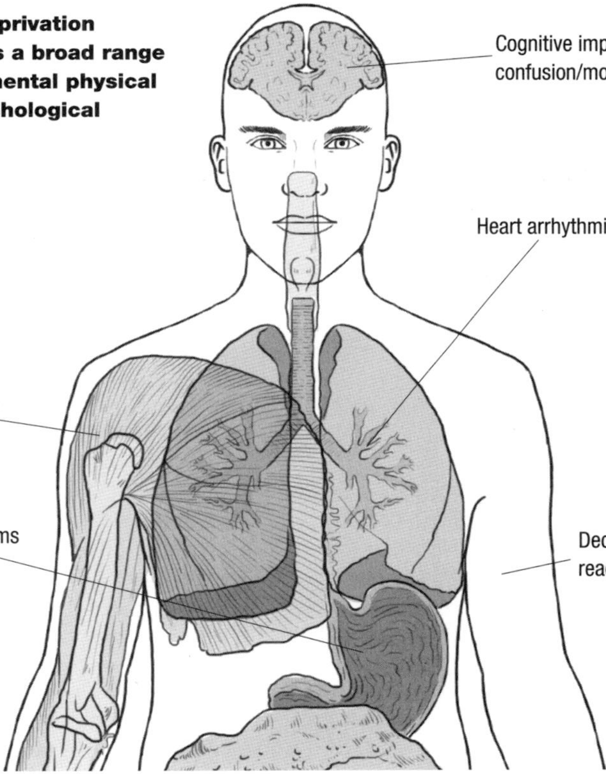

Marine Corps has developed a system for ascertaining which foods can be eaten in unlimited amounts, which foods can be eaten but with moderation and which food should be completely avoided.

They put the system into military speak, labeling these foods respectively "Engage at Will" (eat all you like), "Aimed Shots" (eat in moderation) and "Check Fire" (avoid). A link to information describing which foods come in which categories is provided in the Further Reading on page 188.

However, good nutritional information is easy to come by nowadays. Draw up your own list of foods under these categories, personalizing the list toward foods that you enjoy and will have no problem eating.

Royal Marine Advice about Diet and Nutrition

Your muscles are 75 percent water. Try to stay hydrated before, during and after exercise. Carbohydrates help you recover from exercise and help you run farther and faster. Eat a carbohydrate-rich snack within an hour of finishing your training session, as this is when your muscles need it the most and can store it most efficiently. Foods that are rich in carbohydrates include bread, rice, pasta, cereals and potatoes.

Fat is packed with calories but your body can't use calories as energy for training as easily as carbs. Excess calories will be stored as fat and you will begin to put on weight. Not all fat is bad but you should try to cut down on meat with visible white fat, chips, hard cheese, butter and lard, pies and pastry, cakes and cookies, cream and sour cream. Lean meat, fish, poultry, eggs, nuts and beans are all good sources of protein that your body needs to grow and repair damaged muscles. A chicken or turkey sandwich is just about the perfect post-exercise snack. The bread contains carbs, the meat contains protein, and you'll get all the fat you need from the butter! Alternative sandwiches could be tuna (with a little mayonnaise) or peanut butter. Beans on toast or a glass of milk and a banana would also be good choices.

Try to eat at least five portions of fruit and vegetables a day—more if you can. These can be fresh, frozen, chilled, canned or dried. Pure fruit or vegetable juice also counts, but it doesn't contain as much fiber. And remember that fruit juice is also high in sugar. Milk and dairy products are rich in calcium, which makes your bones and teeth stronger and helps your muscles and nerves work properly. Try to drink a pint of milk a day—2%, 1% or skim milk is best, as it's lower in fat. Smoking affects your fitness levels and immune system, making you more likely to be ill. Try to cut down or give up.

—Royal Marines, *Get Fit to Apply*

Be systematic in both your physical and psychological development, setting clear, structured goals.

Handling Stress

One topic that will underpin much of what we say in this book is about the ways that soldiers are trained to recognize and handle stress. We must not be too negative about the nature of stress because it can be a productive force. Stress is a state of physical and mental tension, fueled by definite changes in the body that make the mind more alert and the body more energetic. It is typically produced by external influences that trigger feelings of anxiety, excitement or fear; in essence, the stress reaction is giving us the energy and focus to deal with the problem most effectively. We all need stress to some degree to help us deliver optimal performance at the required moments.

The problem comes, however, when stress reaches extremely high levels that are sustained for longer than the individual can comfortably endure. When stress reaches critical levels, mental performance deteriorates significantly and the sufferer can either mentally withdraw from the world or show signs of becoming increasingly aggressive and alienated from people and surroundings.

One of the clearest statements of the effects of stress comes from the

Intensive levels of stress, if allowed to continue indefinitely, can lead to mental paralysis and breakdown.

Today we know that the effects of prolonged and negative stress on the body are harmful, affecting everything from sex drive to the immune system.

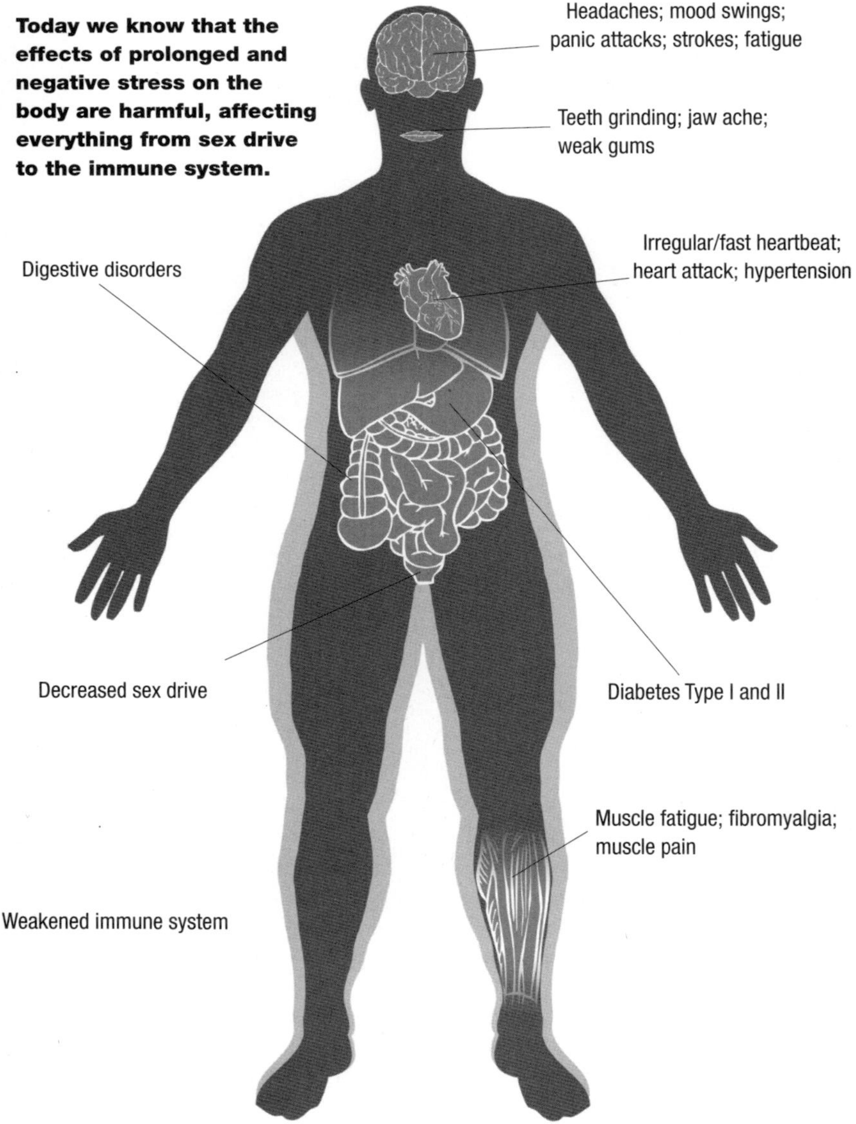

U.S. Army's FM 3-05.70, *Survival*:

We need to have some stress in our lives, but too much of anything can be bad. The goal is to have stress, but not an excess of it. Too much stress can take its toll on people and organizations. Too much stress leads to distress. Distress causes an uncomfortable tension that we try to escape or, preferably, avoid. Listed below are a few of the common signs of distress that you may encounter when faced with too much stress:

- *Difficulty making decisions.*
- *Angry outbursts.*
- *Forgetfulness.*
- *Low energy level.*
- *Constant worrying.*
- *Propensity for mistakes.*
- *Thoughts about death or suicide.*
- *Trouble getting along with others.*
- *Withdrawing from others.*
- *Hiding from responsibilities.*
- *Carelessness.*

—U.S. Army, FM 3-05.70, *Survival*, 2-2

The mind, like the body, needs regular exercise in the form of intellectual pursuits, improving memory, concentration, and logical dexterity.

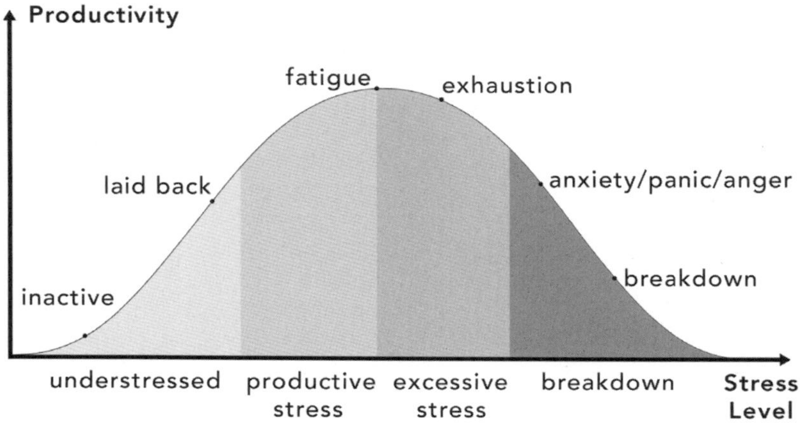

To give this list context, it is important to note that stress is not just a mental state, but also a physiological reaction, which will be discussed in greater depth in subsequent chapters. Long-term stress can result in physical problems such as high blood pressure and cardiac problems, plus risky behaviors (e.g., drug/alcohol abuse) that put the body at further risk, so it needs to be addressed squarely.

Stress and the strategies for dealing with it will appear regularly throughout this book. However, it is worth repeating the point we made at the beginning about perceived spiritual problems often being practical issues. So before you analyze yourself in psychological terms, look at your surroundings to see if there are any "stressors" that can be dealt with on a practical level.

Stress can be productive to a point, delivering energy when needed; but prolonged and intense stress is harmful.

Typical stressors are listed in the U.S. Army's *Combat and Operational Stress Control* manual (see opposite).

The "Environmental" and "Physiological" categories here present a list of stress factors that can be solved largely by practical action. In fact, even many of the "Cognitive" and "Emotional" stressors can be grounded in practical considerations. For example, feelings of isolation and loneliness can be perfectly objective and require the sufferer to make concrete actions toward associating with more people.

A key recommendation of this book, therefore, is that the reader should take a close look at the

List of Stressors from U.S. Army, FM 4-02.51, *Combat and Operational Stress Control*

Physical stressors	Mental stressors
ENVIRONMENTAL Heat, cold, wetness, dust Vibration, noise, blast Noxious odors (fumes, poisons, chemicals) Directed-energy weapons/devices Ionizing radiation Infectious agents Physical work Poor visibility (bright lights, darkness, haze) Difficult or arduous terrain High altitude	COGNITIVE Information (too much or too little) Sensory overload or deprivation Ambiguity, uncertainty, unpredictability Time pressure or waiting Difficult decisions (rules of engagement) Organizational dynamics and changes Hard choices versus no choice Recognition of impaired functioning Working beyond skill level Previous failures
PHYSIOLOGICAL Sleep deprivation Dehydration Malnutrition Poor hygiene Muscular and aerobic fatigue Overuse or underuse of muscles Impaired immune system Illness or injury Sexual frustration Substance use (smoking, caffeine, alcohol) Obesity Poor physical condition	EMOTIONAL Being new in unit, isolated, lonely Fear and anxiety-producing threats (of death, injury, failure or loss) Grief-producing losses (bereavement) Resentment, anger and rage-producing frustration and guilt Inactivity producing boredom Conflicting/divided motives and loyalties Spiritual confrontation or temptation causing loss of faith Interpersonal conflict (unit, buddy) Home-front worries, homesickness Loss of privacy Victimization/harassment Exposure to combat/dead bodies Having to kill

feelings of negative stress in his or her life and be honest about whether the source of that stress relates to the physical body or other manageable factors in the external world. If the stressor can be traced back to a practical problem, then that should be the root of the approach to tackling it. In essence, this is reflecting the famous "Serenity Prayer" written by the American theologian Reinhold Niebuhr:

> *God, grant me the serenity to*
> *accept the things I cannot change,*
> *Courage to change the things I can,*
> *And wisdom to know the*
> *difference.*

Stress becomes a real problem when it affects the full spectrum of your health, causing problems in well-being and behavior.

The Importance of Acting

The final message of this chapter is one that will run throughout many of the discussions in this book, and its importance is critical. For

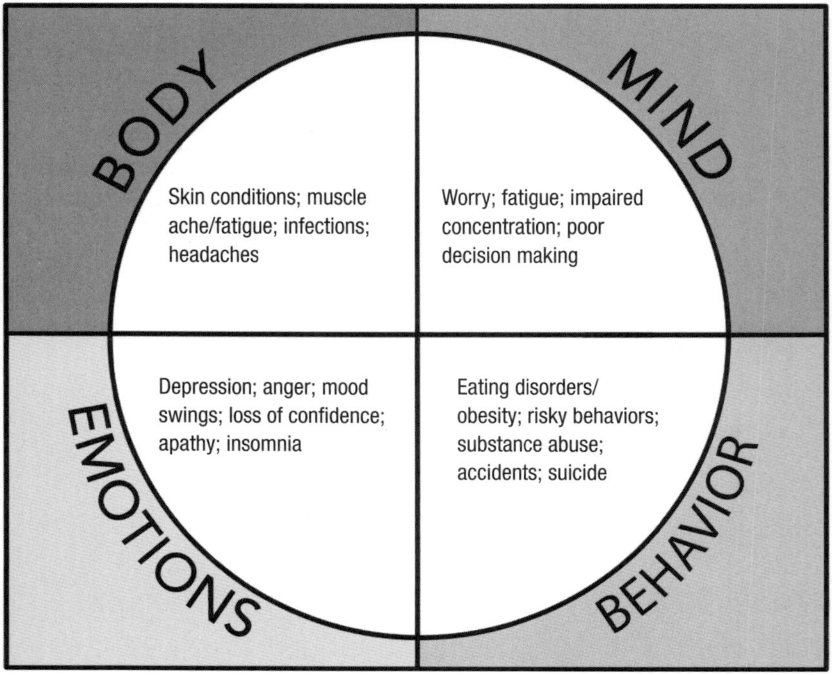

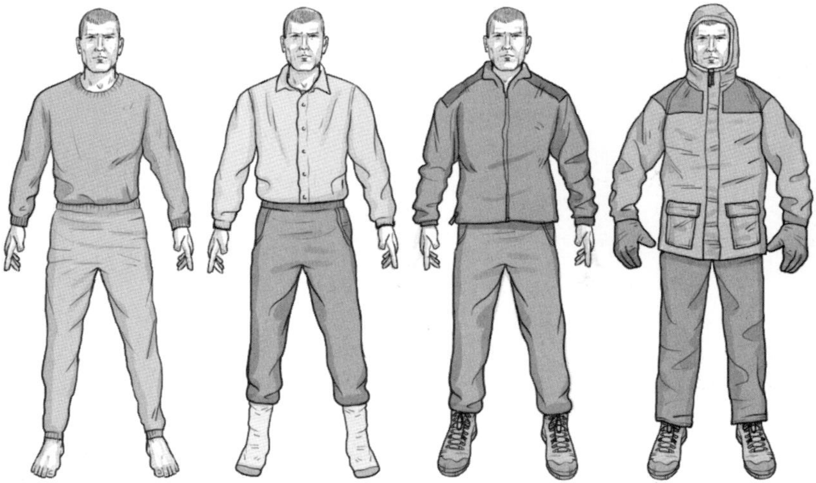

a long period of time in modern psychological understanding, it was the popular and the scientific view that we feel emotions in our mind, then these emotions are expressed through our bodies. So, for example, we first feel depression mentally, that our bodies then follow with the outward expressions of depression—hunched and closed body language; a frowning and pensive expression; lethargic movements; eyes glazed and unfocused.

Yet recent research has shown this model to be inaccurate. Rather, it has been discovered that in many ways our minds take their cues from the body, the mental state conforming to the way that we are physically behaving. This can be demonstrated easily by making a

The right clothing is vital for confidence in harsh physical environments. Here we see a demonstration of the layering principle of heat conservation.

mismatch between our minds and our bodies. For example, if you are feeling depressed, physically push your body into a posture that is the opposite of depression—stand up straight with your shoulders back and chin up; turn your mouth into a confident half smile; keep your hands and arms loose and relaxed; step out in a strong and confident walk. If you do this and persist in maintaining the "act" of someone who is not depressed, then most people typically find that their mental state changes to match their body state. Suddenly, they are no longer depressed because their bodies are telling them so.

The implications of this new understanding are actually quite profound, and in some ways the military has understood this principle for a long time. I was once watching a group of Royal Marine

The harsh training regimens of elite military units are designed to find weaknesses in both mind and body, and replace them with strengths.

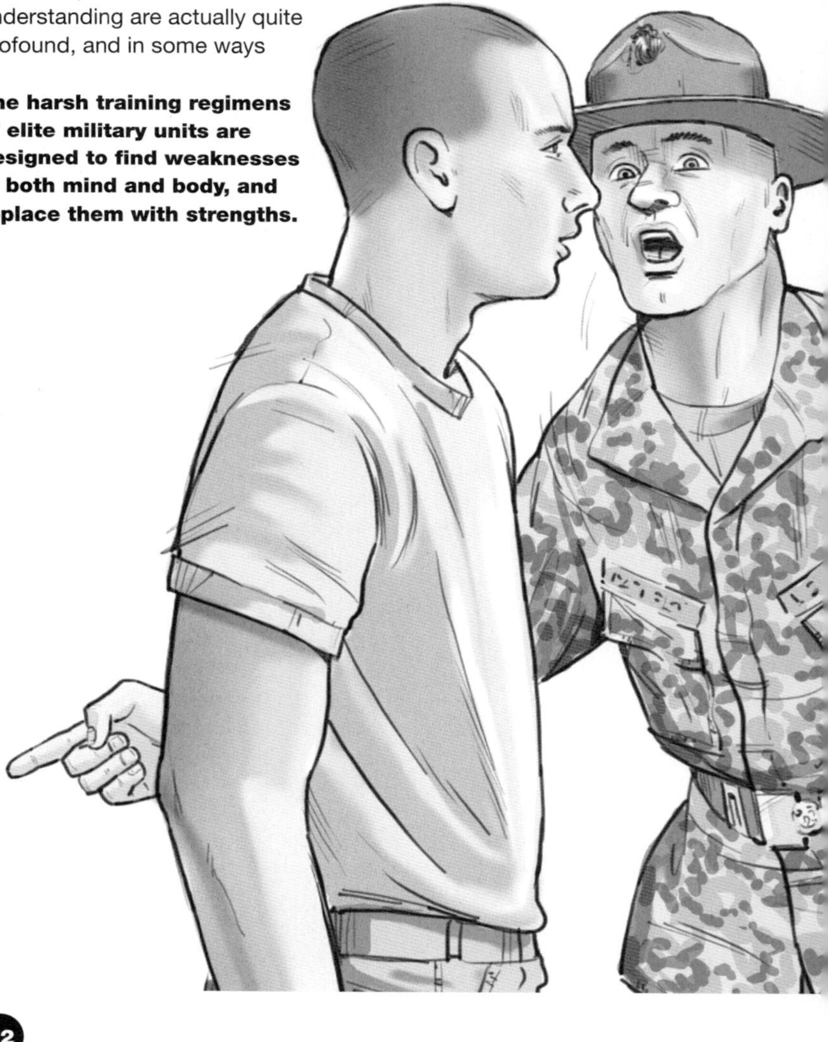

recruits who had just completed an excruciating, lengthy route march. One of the recruits had not made the required time and was told by the instructor that he would have to go back a step in the training program to complete the phase again, leaving his existing training unit and the friends he had made. The young man's face dropped to the floor in dejection, but quick as lightning the instructor told him to raise his chin and keep it there. Effectively, the instructor was forcing the recruit to avoid the body language of defeat, and thereby compel his mind to face up to reality and to engage with it forcefully.

Exuding Confidence

Military forces are grounded in confident body language—standing straight on the parade ground; screaming with animal aggression during unarmed combat and bayonet training; practicing giving clear and firm orders in leadership school; shouting "Oo-rah" (in some U.S. military units) as a gesture of martial spirit. The cumulative effect of such body confidence is that it is easier to breed mental confidence among the military personnel.

The same principle can be applied in our daily lives. If you are faced with a challenge, a problem that seems intractable, first create a picture in your mind of the type of person who could handle such a problem. Really picture him or her—the body language, the tone of voice, the speed at which that person walks, the way he moves his hands while talking, even the clothes he is wearing. Now model your moves on that image, acting the character out on a physical level, although instinctively without mannerisms that are fundamentally unnatural to observers. By acting "as if" you already have the mental state that you desire, you will often find that your mind conforms to the new and more purposeful body language.

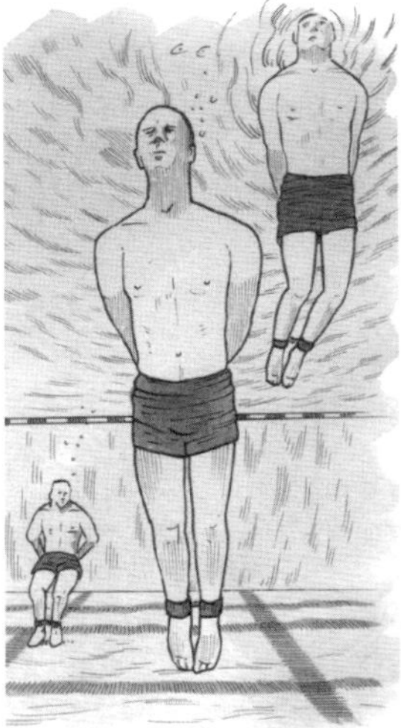

2

Training

Training is the critical transitional phase between civilian and soldier, and the process by which new skills are instilled and old ones refreshed. "Train hard, fight easy" is the classic military rule.

In military circles, there are essentially two forms of training. First there is basic training, in which a new recruit learns the core skills that are fundamental to his branch of service. For example, the basic training of someone who joins the infantry will invariably include physical fitness, barracks and parade-ground discipline, basic rifle handling and marksmanship skills, small-unit tactical skills, and military etiquette. If he can pass such training successfully, then he can enter the ranks of the armed services in possession of a core knowledge shared by all other soldiers in his field. Following basic training, military personnel then typically develop their skill set through specialist training. Sticking with the example of an infantryman, the soldier might go on to receive further instruction as, for example, a sniper, forward air observer, driver, logistics handler, demolitions expert, medic or any of a wide range of other military roles. Through the combination of basic and specialist training, the military is able to create a tactically versatile and mutually supporting force of personnel.

Of course, what constitutes basic and specialist training varies

Opposite: Military training is designed to be extreme, to reveal the recruit's real character.

The recruit has to move into the military mindset, characterized by self-discipline, mental resilience, and a team-oriented attitude.

considerably according to the military unit involved. For those wishing to join elite forces such as the Special Air Service (SAS) or U.S. Navy SEALs (SEa, Air and Land), the "basic" level is very demanding on both physical and intellectual levels, hence both units exhibit a 75–90 percent failure or dropout rate among new recruits attempting to pass the selection process. Indeed, the SAS will only take recruits who have already served in the armed forces and completed a basic training process.

In terms of mental development, what is military training supposed to accomplish? In essence, all military training is designed to achieve the following objectives:

- Transfer the knowledge of general and specific military skills.

- Expose any character flaws in the individual, and either correct those flaws or reject the soldier on the basis of them.
- Give the soldier a degree of conformity with the mindset and aspirations of the wider military unit or formation.
- Foster resilience, self-reliance and initiative.
- Ensure that the soldier can work within the context of a highly integrated team.
- Determine which individuals have leadership skills.
- Ensure that the soldier can handle extreme levels of mental and physical stress.

At its heart, this list boils down to both exposing character and to

A soldier fords a river, keeping his M16 rifle out of the water despite the burning sensation slowly growing within his arm muscles.

building character. These principles are useful, as they can give a discipline and shape to the idea of our own character development.

The Right Attitude

From the outset of basic training, military instructors are looking for certain emotional qualities in the man or woman in front of them. They want to see someone who is tenacious, mentally strong, honest and tolerant, self-disciplined, possessing a good sense of humor (essential to survive what is to come) and a positive attitude, able to get along with people, capable of switching aggression off and on when

Military training has the unique characteristic of instilling controlled aggression, with the ability to unleash violence against an identified enemy.

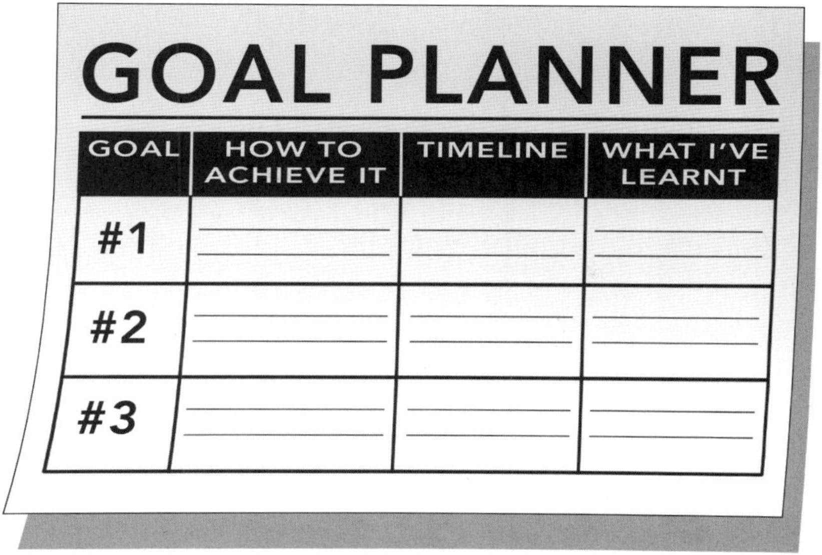

Any effective training program should be sequential and progressive. Use a goal planner to rationalize the schedule and daily/weekly objectives.

required, and also someone who is self-reflective enough to realize when they are doing something wrong and mature enough to resolve the issue. Some of these qualities will be instilled or strengthened through the training, but the recruit must arrive at basic training on Day One already possessing a certain strength of character.

Some of the best psychological advice for those wanting to engage with military training actually comes from the U.S. Army's field manual 3-05.70 *Survival*. In one particularly useful set of guidelines, it explains the right mental state you should bring to a survival situation, but these are equally applicable to military training or indeed to any challenging situation in life:

> *2-26. KNOW YOURSELF*
> *You should take the time through training, family and friends to discover who you are on the inside. Strengthen your stronger qualities and develop the areas that you know are necessary to survive.*

A U.S. Army unit trains in urban assault tactics, repeating tactical patterns of movement until they become second nature.

2-27. ANTICIPATE FEARS
Don't pretend that you will have no fears. Begin thinking about what would frighten you the most if forced to survive alone. Train in those areas of concern to you. The goal is not to eliminate the fear, but to build confidence in your ability to function despite your fears.

2-28. BE REALISTIC
Don't be afraid to make an honest appraisal of situations. See circumstances as they are, not as you want them to be. Keep your hopes and expectations within the estimate of the situation. When you go into a survival setting with unrealistic expectations, you may be laying the groundwork for bitter disappointment. Follow the adage, "Hope for the best, prepare for the worst." It is much easier to adjust to pleasant surprises about your unexpected good fortunes than to be upset by your unexpected harsh circumstances.

2-29. ADOPT A POSITIVE ATTITUDE
Learn to see the potential good in everything. Looking for the good not only boosts morale, it also is excellent for exercising your imagination and creativity.

2-30. REMIND YOURSELF WHAT IS AT STAKE
Failure to prepare yourself psychologically to cope with

survival leads to reactions such as depression, carelessness, inattention, loss of confidence, poor decision making and giving up before the body gives in. Remember that your life and the lives of others who depend on you are at stake.

2-31. TRAIN
Through military training and life experiences, begin today to prepare yourself to cope with the rigors of survival. Demonstrating your skills in training will give you the confidence to call upon them should the need arise. Remember, the more realistic the training, the less overwhelming an actual survival setting will be.

—U.S. Army, FM 3-05.70, *Survival*, 2-26–2-31

In parachute training, natural fear needs to be controlled by a focus on logical and disciplined attention to safety procedures.

Be Realistic

There is much in this list that warrants unpacking in further detail, and it can form the core of our understanding of "basic" mental training. I start with the principle "Be realistic." Soldiers have to be ruthlessly honest about the situation confronting them, for there is no wishful thinking in combat, and any underestimations of the enemy or the environment will be ruthlessly exposed (see the next chapter). When facing a particularly challenging situation, the soldier must first look at it with grim honesty and assess the nature of the threat or demand. Then, equally honestly, he must define the actions required to rectify the problem or meet the

The advent of 3D game imaging raises the possibility of ultra-realistic "virtual" training, the complete immersion in the combat experience.

challenge, including the intellectual skills or physical fitness necessary.

As a way of coping with complex situations, a soldier should break down the challenge into small, manageable objectives and consciously design the practical steps to deal with each of those objectives in turn. Most critical of all, he must implement the plan vigorously in the real world, setting himself targets with demanding but achievable deadlines. Remember—a plan in itself means nothing without action, in the same way that talking about what you aim to achieve is just hot air without meaningful activity. A soldier must orient his life to being a doer, not a talker.

Positive Outlook

There can be much misunderstanding about what constitutes a positive attitude. Given what we have just said about realism, a positive attitude must not be confused with Pollyanna thinking, denying hard truths in favor of dreamy thoughts. Instead, a positive attitude is more about looking at the world with humor and with a sense of opportunity. Right now, look around the room or surroundings you are in. I want you to identify five items, features or even people in your environment that satisfy one or more of the following criteria:

- Something that is physically beautiful.

Trainers look for clear signs of cooperative problem solving in recruits. Here a team of three recruits masters the wall climb.

- Something that presents an opportunity to develop your knowledge.
- Something that offers a possible solution to a problem.
- Something that evokes a nice memory.
- Something that gives you an idea for a new venture or hobby.

Effective training is about pace management, being active and focused when it is required, but switching off and relaxing between the bursts of activity.

The more a soldier is exposed to the real physical sensations of combat in training, such as the shock wave from this explosion, the better he will cope with combat for real.

It may be that if your surroundings are particularly grim, you might struggle to find objects that satisfy all of these criteria. Yet notice how different the world looks, and also how different you feel, when you actively scan your environment for stimulus that encourages you to feel positive and in control. So, as part of your "basic" mental training, approach every situation with your eyes open for productive opportunities and beautiful sights, as doing so will improve your core resilience.

Visualization and Training

In the passage quoted from the *Survival* manual above, the last section explained: "Through military training and life experiences, begin today to prepare yourself to cope with the rigors of survival." An additional tool for preparation, one that is either unwittingly or explicitly used in many military training regimens, is visualization. This is the technique of imagining something so vividly in a meditative state that it acts as a form of life experience, giving the mind a resource to draw upon should the individual actually face the situation in real life. Put in rather crude terms, the human mind operates partly like a large document folder on a computer. Every time you have a new experience, a new file is created in the folder. When you repeat the experience, in essence you go back to the folder, retrieve the relevant file and add data to it. This is how, through training, you become better and better at a particular skill, as the file contains ever-expanding information about how to do something more competently. When we are faced with a completely unfamiliar situation, however, there is little in our folder of experiences to draw upon. Hence we attempt to adapt an existing experience to the new one, in the process creating a new file for the novel experience.

Taking this model, military trainers are faced with the particular

challenge of how to train soldiers to face the traumatic realities of combat. Obviously, at a training level the soldiers cannot engage in actual combat, so the solution is to create a training environment that is as faithful to actual combat as it can be without real casualties being taken. (It is noteworthy that in elite units there can be a 5–10 percent casualty rate during training, indicative of the realism involved.) In some of the most advanced training, measures to achieve this include:

Shared adversity in training, backed by a robust sense of humor, helps breed the esprit de corps that forms the foundation of a unit's morale.

- The use of live ammunition to acclimatize the soldiers to actual danger and the noises of combat.
- The use of "simunitions"—non-lethal but extremely painful rounds that can be fired directly against "opponents."
- Advanced make-up and prosthetic techniques to create

highly realistic combat wounds to acclimatize the soldiers to the reality of violence.
- Reconstructed towns and villages to familiarize the soldiers with combat environments.
- High-pressure training scenarios in which the trainees are put through extreme mental demands to get them used to coping with adrenaline and pressure.

The lesson for those in civilian life is that the most productive training for any challenge is that which most closely replicates conditions you will face. An essential component of mental toughness is to face situations that we find uncomfortable or challenging, and by coping with them, building up our psychological "muscle." Essentially, through hard training we condition ourselves to cope.

Visualization is another tool we can use to achieve this conditioning, although ideally it should only be an accessory to practical training. Visualization works by the practitioner's imagining the way a specific situation is handled. If imagined in vivid detail, the mind can actually form a new file for dealing with that experience without having encountered it practically. Visualization takes some technique and practice to master, but here are its basic principles:

When exhaustion sets in, imagine that you are being powered by a large engine located in the chest, driving your legs forward with each step.

1. Find a quiet place where you can lie down undisturbed.

2. Now focus on the rise and fall of your breathing. With every out breath, feel all the muscles in your body soften and relax. Imagine that your body is like a pat of butter on a summer morning windowsill, gently softening under the warming sun. If any unwanted thoughts move into your head, don't attach any emotion to them—simply observe them and let them just drift away, as if you're watching them pass fleetingly across a TV screen. Return your attention to your breathing.

3. After about five minutes of deep relaxation, now create a picture in your mind of the situation you want to deal with. The important point is to imagine it with absolute realism. Mentally re-create all the sights, sounds, smells, sensations and emotions involved. The more vivid the picture you create, the more mentally beneficial it will be.

4. Now imagine being in the situation behaving in the manner you want to achieve. For example, if you are facing a major public presentation,

imagine being in that presentation with a confident voice and demeanor, and a balanced and relaxed posture. See the audience responding to you and interacting enthusiastically. Feel your words flowing out creatively. Remember the "acting out" principle described at the beginning of this book. If you don't know what it might feel like to behave in the new way, mentally model the behavior of someone you know to be a confident public speaker. In many ways, you will be doing what soldiers do when they model their own behavior on that of a particularly inspirational leader.

5. One psychological trick to improve the post-meditation effect of visualization is to create a physical anchor to the imagined behavior. At a key point in your visualization, when

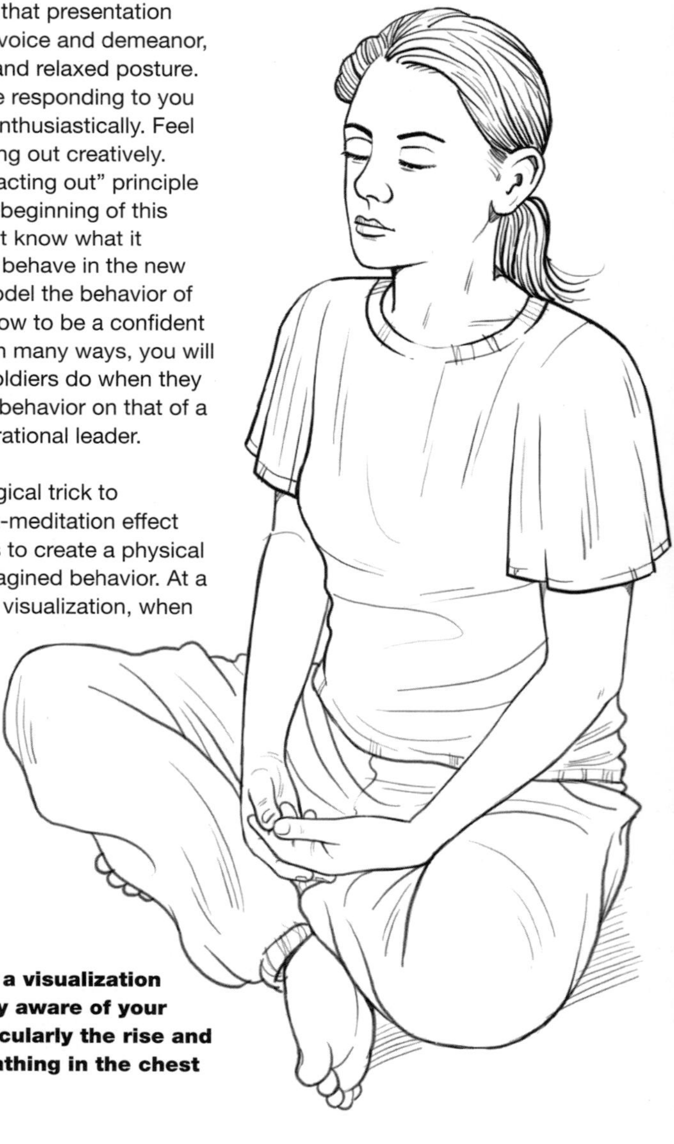

When you start a visualization session, be very aware of your own body, particularly the rise and fall of your breathing in the chest and abdomen.

you feel that the mental behavior you have created is almost exactly as you wanted, pinch the thumb and forefinger of your right hand together to create a link between body and behavior. After your visualization session, by pinching your thumb and forefinger together again you can invoke the feelings and impressions that you created during the visualization.

6. Once you have mentally "rehearsed" the situation thoroughly, take five deep breaths in and out, making your body more wakeful with

Moving visualization: Imagine a tight bungee cord attached between you and your objective, the cord's tension drawing you effortlessly onward.

each breath. Then open your eyes and finish the session.

Practice your visualization routine several times on the same scenario, each time improving the quality of the "picture" and the strength of your desired behaviors in it. Note also that visualization can be used on a more improvised and immediate basis to help increase your performance. Simply picturing yourself doing something well and achieving the desired outcome just before you actually do it can improve the real-world results.

U.S. Navy SEAL's "drownproofing" —with wrists and ankles bound— is designed to test the recruits' ability to handle rising panic.

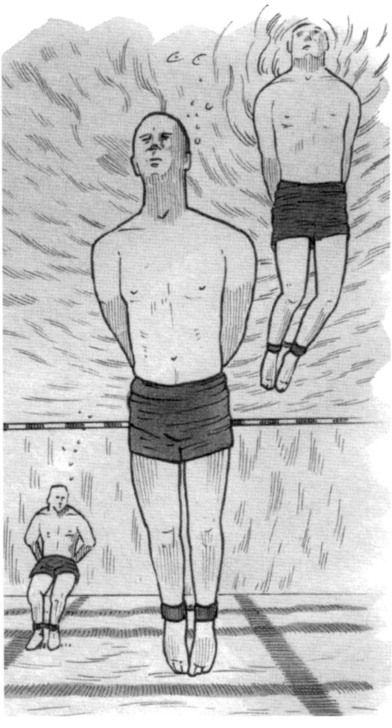

Mental Agility

Visualization, experience, and conditioning are the fundamental building blocks of your mental "basic training." We now need to reflect more on how to develop specialist skills and abilities.

One characteristic of elite Special Forces units tends to be an extremely high investment in the skill set of each individual. By sinking a lot of time and money in the knowledge spectrum of the unit, Special Forces end up with extremely flexible and self-reliant people. The aforementioned U.S. Marine Corps *Warfighting* field manual devotes considerable space to thinking about the issue of military education. It sees education not as an optional intellectual pursuit in addition to the hard business of soldiering, but as integral to it—particularly in the creation of effective leaders:

> *The early stages of a leader's career are, in effect, an apprenticeship. While receiving a foundation in theory and concepts that will serve them throughout*

their careers, leaders focus on understanding the requirements and learning and applying the procedures and techniques associated with a particular field. This is when they learn their trades as aviators, infantrymen, artillerymen or logisticians. As they progress, leaders should strive to master their respective fields and to understand the interrelationship of the techniques and procedures within the field.

A Marine's goal at this stage is to become an expert in the tactical level of war. As an officer continues to develop, mastery should encompass a broader range of subjects and should extend to the operational level of war. At this stage, an officer should not only be an expert in tactics and techniques but should also understand combined arms, amphibious warfare and expeditionary operations. At the senior levels, an officer should be fully capable of articulating, applying and integrating MAGTF [Marine Air-Ground Task Force] warfighting capabilities in a joint and multinational environment and should be an expert in the art of war at all levels. [...]

Military training also incorporates ceremony and symbolism, using both to raise the soldier's sense of self-worth and pride in his unit.

> *Finally, every Marine has an individual responsibility to study the profession of arms. A leader without either interest in or knowledge of the history and theory of warfare—the intellectual content of the military profession—is a leader in appearance only. Self-directed study in the art and science of war is at least equal in importance to maintaining physical condition and should receive at least equal time. This is particularly true among officers; after all, the mind is an officer's principal weapon.*
>
> – U.S. Marine Corps, MCDP 1, *Warfighting*, p.62

This passage makes some important points about the intellectual development of the self. Not least, it advocates not just becoming informed about something, but becoming an expert in that topic, through formal study and also through self-education. This message is particularly relevant for our media-saturated age, in which a massive overload of diffuse information detracts from the focus required to achieve expertise. Based on how the professional military educates its leaders and soldiers, try to build the following into your mental fitness regimen:

Time and Practice

Seek to become an expert in at least one subject area. Achieving expertise requires considerable disciplined time input—experts estimate that to reach true expert level requires 10,000 hours of practice in a particular discipline. Don't be dissuaded by such a hefty commitment. Instead, define what you want to achieve in which subject area, and then allocate a portion of every day to reading about or researching this topic. Try to ensure that the learning contains some active or social element (such

as joining a discussion group), as learning is accelerated if the topic is practiced in multiple learning styles (spoken, written, reading and/or listening).

Read Well

To become an expert on a topic, there is no substitute for spending a lot of time reading quality publications. I emphasize the word "quality." There is much poor information on the Internet (and indeed in some books), which can breed ignorance more than knowledge. Furthermore, the very nature of the Internet encourages jumping from topic to topic in an erratic fashion, leading to distraction and the poor retention of information. Therefore it is always best to select high-quality publications, or at least detailed and academically approved websites.

Inject some discipline into your reading to get the most out of it, making it "active" rather than "passive." Before you start your reading session, first determine some guiding questions about what you will look for in the narrative. For example, if you are going to read about the origins of World War II, you could ask yourself before reading: "In what ways did the U.S.'s foreign policy contribute toward the outbreak of the war?" Although much of the information you read may not relate to this question specifically, the fact that you are looking for the answer will make your mind more sensitive to the content that you are absorbing and will aid in its retention.

During your reading, mentally interact with the text constantly, asking questions, pausing to think about implications, making parallels with your own life and visually imagining the scenes described—

Mapwork from a tank commander's course. The technicalities of military service demand intellectual training as intense as the physical training.

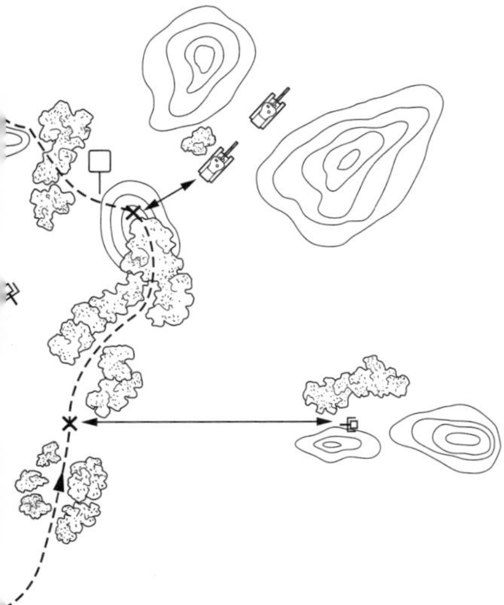

Mechanical tasks, such as the all-important duty of dismantling and cleaning a rifle, should be performed with such frequency that they embed themselves in muscle memory and become almost second nature.

do anything to force a high-level of interaction with a text, which again helps you to retain what you have read. Also, keep your eyes open to what you consider the most essential facts in the narrative; mentally "spear" these facts into your mind and use them to hang the rest of the factual content upon.

Finally, include an element of review in your reading. When you pick up the book again after a day's break, skim over the content you read previously, quickly refreshing the essential facts and reminding yourself of some of the key elements. Following this overall approach to reading will mean that you increase your retention and thereby expand your repository of usable knowledge.

Increased Memory Power

Memory is fundamental to intellectual power, and yet it is one of the qualities that are most under threat nowadays. Much research indicates that the fragmented nature of learning that comes from excessive Internet use is not conducive to memory, as the user ultimately comes to rely upon the device rather than knowledge to answer a question.

Good military training always starts from fundamental principles rather than technological solutions. For example, although many soldiers will ultimately come to rely upon GPS navigation systems in the field, the military still takes pains to teach

At all times during training, stay properly and consciously hydrated, to ensure your best bodily performance and to avoid cognitive impairment.

every soldier traditional methods of map and compass reading. Not only does this provide a soldier with a fail-safe backup should technology fail, but it also stretches the soldier intellectually and builds a new skill into his memory. There are numerous ways of increasing memory power, and entire books are written upon this very topic. The following are some simple ways to boost memory:

Memory exercise: Try to connect each image through visually imaginative linkages, to memorize the sequence.

- **Association**—associate items in a list with a connected and unusual mental image.

An example of a memory palace. Certain rooms, represented on the plan, are associated with certain themes—"go" to the room to "see" the memory.

For example, if you wanted to remember five items in a supermarket—bread, carrots, chocolate cake, dog food, and cookies—it is easy to remember them by linking them in the following way: a piece of bread is being stabbed violently right through with a carrot; the tip of the carrot hits a chocolate cake beneath and sprays chocolate everywhere; the greedy dog drops the cookie he is chewing to lap greedily at the chocolate as it splatters his face. The unusual nature of the image is deliberate—the more outlandish or extreme the visual representation and the connection between the objects, the more easily it is remembered.

- **Memory palace**—this classic memory technique involves sorting the individual items that need to be remembered into individual rooms within an imagined house or building. For example, if you intended to be an expert in the history of warfare, you could earmark a separate room for World War II. In that room, there could be a chest of drawers, and each draw is labeled with a different campaign in the conflict. Open a drawer, and

For soldiers, the true purpose of their lengthy training process comes alive on actual deployment to a combat zone.

inside you will "see" words and images that remind you of the facts relating to that campaign. Again, the memory palace is a technique of ordering your information mentally, rather than just relying upon abstract recall of words and dates.

- **Date hooks**—remembering dates and numbers can always be problematic because of their fundamentally unmemorable nature. There are some techniques, however, that can help you anchor dates more firmly in your mind. One of the most easy to use is simply to drop the first two digits of the century date, then create a rhyming aide memoir to help you remember the second two digits. So, for example, if you wanted to remember the starting year of World War II (1939), then drop the "19," and remember "39" through the words "squirty mine." To connect the words to the concept, imagine German soldiers marching into Poland but treading on a land mine that, instead of exploding, squirts out cream. Again, don't be put off by the very abstract nature of the imagery—the stranger the better.

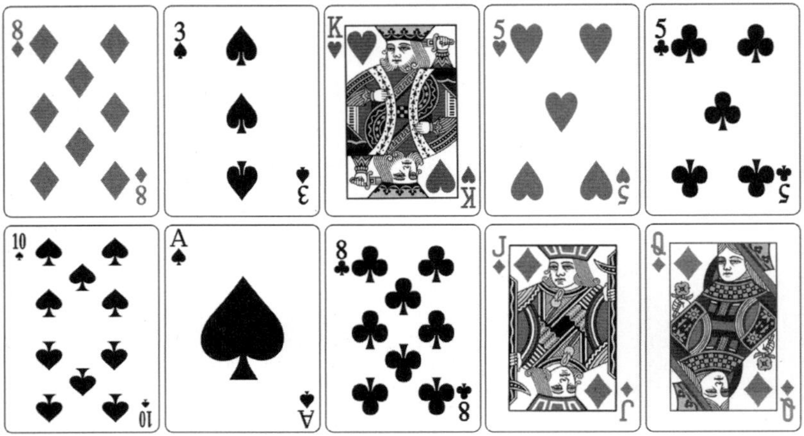

- **Learn a language**—studying a new language not only offers practical benefits for communication in foreign countries, it also provides a superb workout for the brain in terms of memory and logic. Professional military forces now invest heavily in language training for many of its soldiers, as they have realized that possessing language skills gives its forces an operational and intelligence advantage. Many high-achieving military leaders have been polyglots. General Joseph Stilwell, the commanding general of U.S. forces in the China-Burma-India (CBI) Theater during World War II, spoke Spanish, Japanese, and Chinese, diligently picking up new languages in whichever theater he was deployed. Having these language skills brought him unique access to levels of foreign government military command that ordinarily would have been denied or limited to him.

Simple memory exercises, such as remembering the sequence of a deck of cards, reinforce the ability to absorb and process information rapidly.

- **Mnemonics**—these are mental devices of various types that help us to remember something. The easiest sort of mnemonic is an abbreviation. So, for example, in the military, Air Defense Artillery is represented by the abbreviation ADA, which can be said as a single word for convenient memorizing, essentially becoming an acronym. If you have a list of concepts or names to remember, try to create an abbreviation or acronym to act as prompts to remember these names. If you need to make visual anchor points for this abbreviation, you can create a picture narrative in your mind by using the words associated with the military phonetic alphabet (see the feature box).

Leadership

While some men and women join the military at the lowest levels of the rank hierarchy, others attempt to obtain a commission as an officer, or rise through the ranks to obtain officer level. Those who aspire to be officers obviously have to display one quality above all others—leadership.

The key qualities of good leadership can be quite intangible. Some people just naturally seem to project the leader personality, while others seem more destined for a supporting role. (This is not to imply any criticism of those who follow rather than lead, as the only way that the vision of leaders can be realized is through intelligent and effective supporters.)

The U.S. Army has invested much in defining what it is that makes a leader. It makes the distinction between the "Attributes" of a leader—essentially what a leader is—and "Core Leader Competencies," or what a leader

NATO Phonetic Alphabet

A = Alpha
B = Bravo
C = Charlie
D = Delta
E = Echo
F = Foxtrot
G = Golf
H = Hotel
I = India
J = Juliet
K = Kilo
L = Lima
M = Mike
N = November
O = Oscar
P = Papa
Q = Québec
R = Romeo
S = Sierra
T = Tango
U = Uniform
V = Victor
W = Whiskey
X = X-ray
Y = Yankee
Z = Zulu

does. Basically, under this model a leader is defined by who he is and what he does. The army leadership manual provides more detail:

> ***Attributes***
> *What an Army leader is:*
>
> *A leader of character*
> - *Army values*
> - *Empathy*
> - *Warrior ethos*
>
> *A leader with presence*
> - *Military bearing*
> - *Physically fit*
> - *Composed, confident*
> - *Resilient*
>
> *A leader with intellectual capacity*
> - *Mental agility*
> - *Sound judgment*
> - *Innovation*
> - *Interpersonal tact*
> - *Domain knowledge*
>
> ***Core Leader Competencies***
> *What an Army leader does:*
>
> *Leads*
> - *Leads others*
> - *Extensive influence beyond the chain of command*
> - *Leads by example*
> - *Communicates*
>
> *Develops*
> - *Creates a positive environment*
> - *Prepares self*
> - *Develops others*
>
> *Achieves*
> - *Gets results*
>
> – U.S. Army, FM 6-22, *Leader Development*, 2-4

Left: A uniform is an important element in adopting the military mindset. Never underestimate the impact of outward impressions.

Here we see a long and demanding list of qualities and competencies. It is clear from this list that a good leader not only must possess a strong character, one that acts as an inspiration to his subordinates, but he must also have an unswerving bias for action and for getting jobs accomplished.

Space here does not allow us to unpack each of these qualities in turn; reading the respective U.S.

A U.S. Army structure of leadership development. Note how the structure sits upon elements of moral character.

During group exercises, drill officers will look for those who encourage others to succeed in their challenges, a key ingredient of leadership qualities.

field manual will give you deeper information about the development of leadership qualities. Here, however, it is worth noting some of the elements that are frequently misunderstood about leadership and its development.

Good Leaders Listen, Good Leaders Delegate

If you are adopting a military leadership role, do not think that means that you override others' opinions and rely solely on your own judgment. Good officers and NCOs (non-commissioned officers), particularly those in elite and professional units, actively encourage criticism and feedback on their decisions, and also intensively gather expert knowledge from others before making a decision. They do not simply think that they know best.

What they do understand, however, is that they must make a decision and issue a command at the end of the information-gathering process, even if there are still dissenting opinions. Both during and after their decision, they must also delegate authority to competent individuals for the completion of

tasks. A leader who attempts to do everything alone will typically end up overstretched and inefficient.

The Power of Influence

In military teams, the behavior of leaders is absolutely critical in spreading confidence, or otherwise, among the rest of the unit. Emotions are contagious, which is why a soldier who descends into traumatic collapse on the frontline is quickly whisked away to the rear, not just for his own treatment but also to stop his anxiety from spreading to others. Similarly, a leader must project calmness, confidence, reassurance

A U.S. Marine designated marksman (DM). The sniping role requires high levels of self-discipline and tactical awareness.

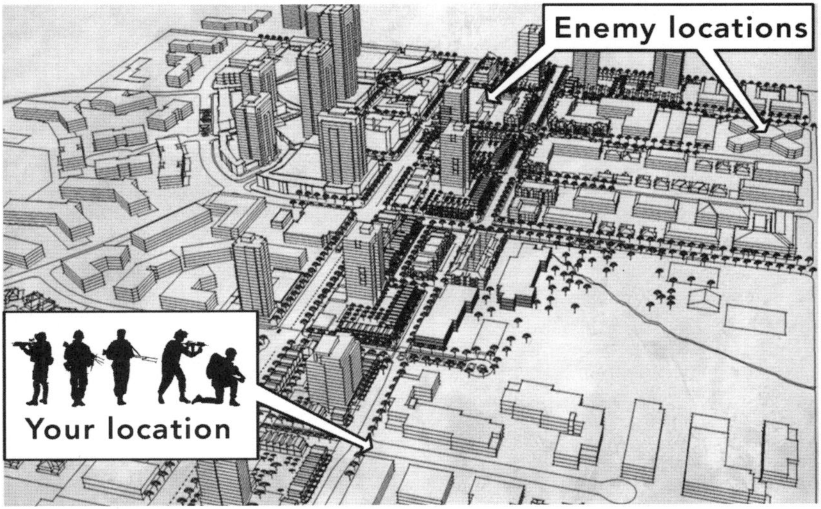

Tactical thinking requires a leader to visualize routes of maneuver around and toward the enemy, while allowing for the inevitable frustration of plans.

and the expectation of victory in all his mannerisms and decisions, even if internally he still feels the anxiety and responsibility of his position. This is where the "acting" principle outlined above can come in useful; the leader can self-consciously put on the character of a leader to project leadership qualities to his unit.

Act Swiftly

As we shall see in the next chapter, combat is a particularly unforgiving environment for decision makers. It takes place rapidly, often brutally, and constantly evolves in new directions. In such circumstances, leaders must be capable of making rapid decisions. The decision will be made, in many cases, on the basis of imperfect information, but making no decision can be worse than making an imperfect decision.

One technique used by some military forces is to provide three (no more, no less) possible solutions to any problem. Then the commander has to choose the option that has the best chance of working. Note that I say "best chance." There are rarely 100 percent guaranteed decisions in combat, so the commander must go for the option with the best chance of success, and rely upon the initiative of his soldiers to work out the kinks as they go.

Communicate Clearly

Leaders must be excellent communicators, but this does not mean that they talk in advanced academic language. Instead, military commands must be issued in plain and easily comprehensible statements (short, emphatic sentences with no confusing terminology) that transfer a clear mission objective. If everyone understands the mission objective, then the individual soldiers have more opportunity to use their initiative to obtain the objective on behalf of their leader.

Although not everyone's character is ideally suited to leadership positions, leadership qualities can

Explosive Ordnance Disposal (EOD) soldiers require unflappable characters, focusing on the task rather than the dangers.

be inculcated through adopting and practicing the role, learning constantly from those around you. As part of its leadership development process, the U.S. military encourages self-analysis as a way of identifying personal strengths and weaknesses, fostering the former and dealing with the latter. Part of this self-analysis test is included here and is worth conducting as a general assessment of character. "Know yourself" is the precursor to development as a leader.

Complete a Self-analysis

Consider the following items and be as specific as possible. Use the items as necessary to identify unique aspects of personal strengths and developmental needs.

Strengths

The skill or ability at which I am best is ________.

The personal quality that I rely on most for my success is ________.

I am most knowledgeable about ________.

The activities I look forward to include ________.

I would love to learn more about ________.

The accomplishment I am most proud of is ________.

Others usually come to me for help with ________.

Others think the best position for me would be ________.

Developmental Needs

The skill or ability that is always difficult for me is ________.

I don't know as much as I should about ________.

I usually go to others for help on ________.

The situation that causes me the most frustration is ________.

I am most hesitant when I try to ________.

I am most concerned about my ________.

Others think I am not skilled at ________.

I would become a more valued member of my organization if I ________.

—U.S. Army, FM 6-22, *Leader Development*, 4-5

3

Engaging the Enemy

Confronting a violent enemy is the ultimate human experience. It is a moment of truth for many soldiers, putting intense pressure on their characters and their training.

An army that maintains its cohesion under the most murderous fire; that cannot be shaken by imaginary fears and resists well-founded ones with all its might; that, proud of its victories, will not lose the strength to obey orders and its respect and trust for its officers even in defeat; whose physical power, like the muscles of an athlete, has been steeled by training in privation and effort; a force that regards such efforts as a means to victory rather than a curse on its cause; that is mindful of all these duties and qualities by virtue of the single powerful idea of the honor of its arms—such an army is imbued with the true military spirit.

—Carl von Clausewitz, *On War*, p.187

It is a common misunderstanding in the civilian world that soldiers only reluctantly embrace the experience of combat. For most soldiers, particularly those who have volunteered for military service rather than have been conscripted, combat is the experience they most ardently desire, if only as proof that they have truly acquired their warrior skills. Once combat has been experienced, and repeatedly, the fervor tends to diminish, although some soldiers

Opposite: Warfighting is diverse in nature, from hand-to-hand combat to long-range killing.

Battlefield leadership is a lonely business. The leader needs personal courage, as well as tactical courage to send others into harm's way.

become hooked on the adrenaline of action and struggle to return to more peaceable living.

The opening quotation from the seminal Prussian military thinker Carl von Clausewitz powerfully conveys what he terms as the "military spirit." Some of these factors we will discuss in this chapter, which focuses on the mental qualities necessary for combat and the domination of an external enemy.

Controlled Aggression

It cannot be denied that combat is ultimately about violence, and violence can rarely be a non-emotional event. To close with another human being and kill him is a profound experience, although the responses to that experience can be variable (see the final chapter). What the military seeks to do with its young men and women is to inculcate *controlled aggression*, which is the aggressive instinct needed to do violence, but channeled toward a purposeful objective. Only if a soldier can control his aggression can he be of use to a modern military force.

When aggression is not controlled, war crimes can be a consequence. For example, on March 16, 1968, U.S. Army soldiers from Company C, 1st Battalion, 20th Infantry Regiment, 11th Brigade, 23rd (Americal) Infantry Division arrived in the South Vietnamese hamlet of My Lai. The unit, harried by several months of combat losses and weakened by

Modern armies now have unprecedented levels of battlefield intelligence. Although this is useful, it can reach overwhelming levels.

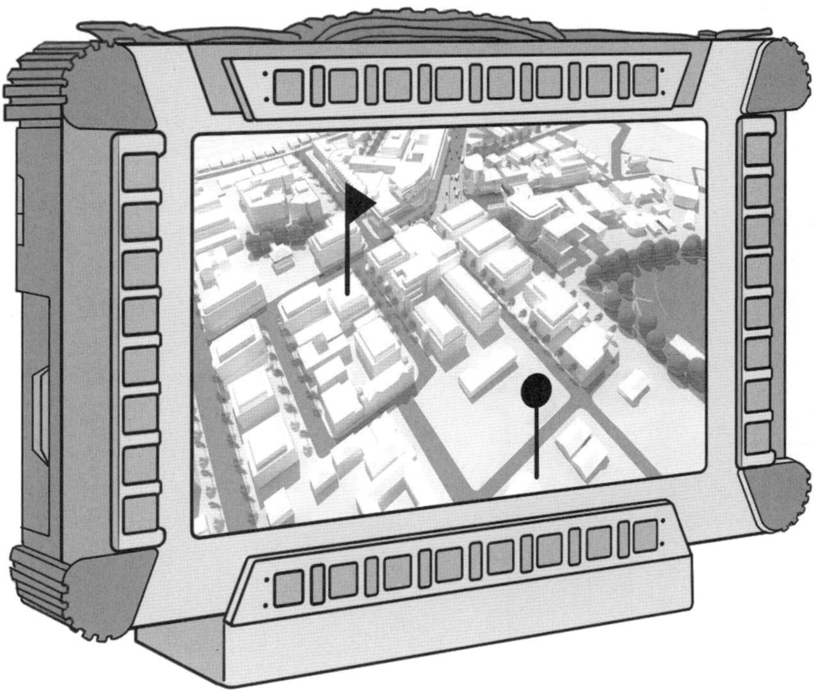

Left: A sniper/ spotter team. The spotter is as much a part of the final shot as the sniper, providing key information to secure the hit. A cool head is needed for both roles.

Bayonet drill is one of the most elemental forms of combat training. Technique is important, but it is secondary to sheer unbridled aggression.

poor leadership, descended upon the civilian population of My Lai with a horrific and unrestrained appetite for violence. Over a period of several hours, possibly as many as 504 men, women and children were massacred in the most brutal fashion. The atrocity, when it emerged in the world press, shocked American society to the core. The psychological causes of this war crime are complex, but one key cause was that the aggression and desire for vengeance that had built up in the soldiers over previous weeks was given free rein through a weak and complicit leadership.

Combat will never be free from dark moments, because violent environments are undeniably breeding grounds for the worst in human behavior. And yet, the best military forces deploy troops who are able to make the switch between furious firepower and humanitarian assistance within the space of minutes. How they are able to do so provides some valuable insights into the general principles of using our adrenaline to achieve results.

So how would we define controlled aggression? A useful inroad into this subject comes from a 1943 U.S. Army publication called, quite simply, *Bayonet*. An introductory passage from the book entitled "The Spirit of the Bayonet" explains some core principles about the nature of controlled aggression:

The will to meet and destroy the enemy in hand-to-hand combat is the spirit of the bayonet. It springs from the fighter's confidence, courage and grim determination, and is the result of vigorous training. Through training, the fighting instinct of the individual soldier is developed to the highest point. The will to use the bayonet first appears in the trainee when he begins to handle it with facility, and increases as his confidence grows. The full development of his physical prowess and complete confidence

Right: Preparation phase. Good soldiers will spend much time diligently servicing and maintaining their weaponry to avoid its critical failure in combat.

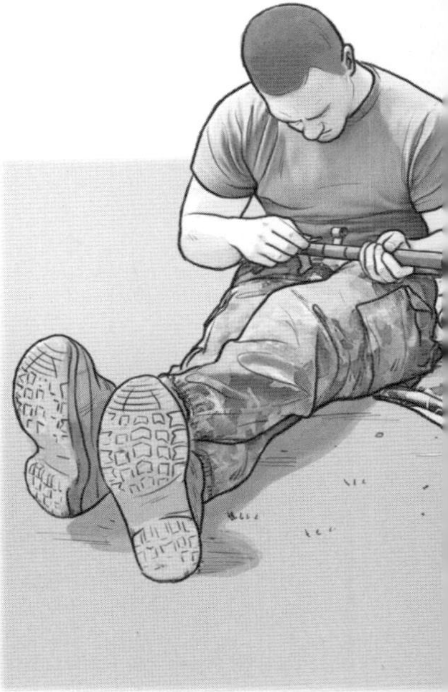

in his weapon culminates in the final expression of the spirit of the bayonet—fierce and relentless destruction of the enemy. For the enemy, demoralizing fear of the bayonet is added to the destructive power of every bomb, shell, bullet and grenade which supports and precedes the bayonet attack.

—U.S. Army, FM 23–25, *Bayonet,* p.1

Above: Soldiers might have to move fluidly between lethal and non-lethal options, as illustrated by these soldiers in the riot-control role.

At the heart of this passage is that close synergy between training and "fighting instinct." It is only through gaining the confidence to handle the weapon well that the soldier is able to embrace the potential of that device and handle it intelligently. Looked at from this perspective, the training does not soften the soldier's aggression, but actually allows it to emerge.

Even though this manual was written in 1943, it in many ways

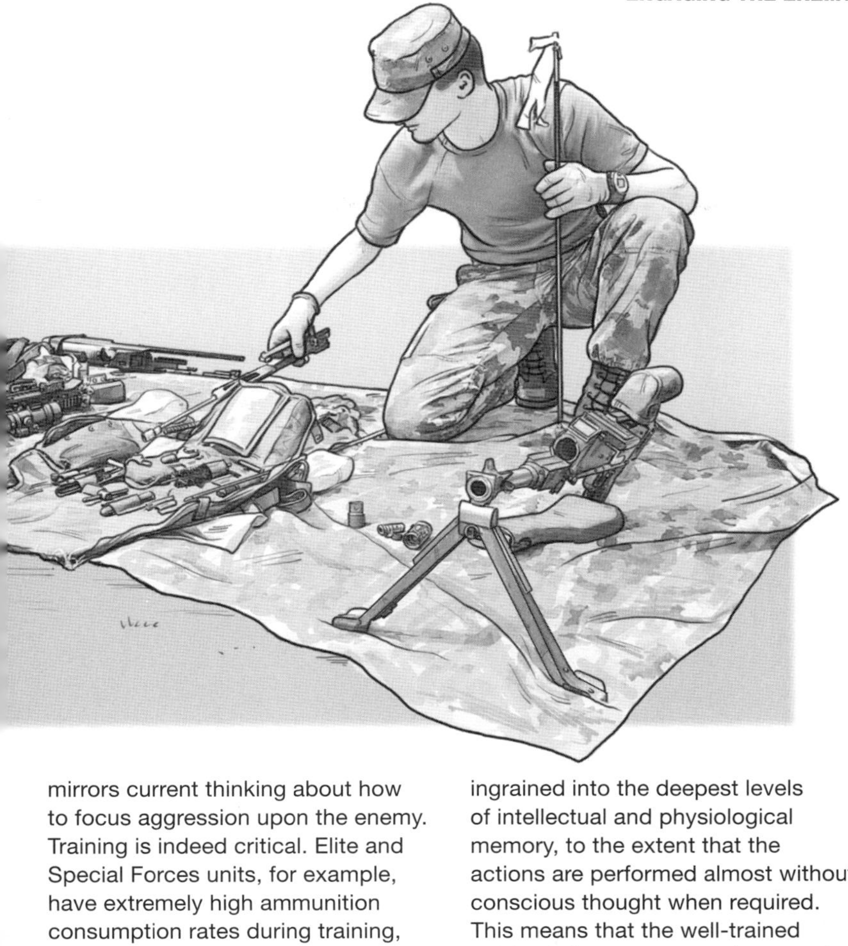

mirrors current thinking about how to focus aggression upon the enemy. Training is indeed critical. Elite and Special Forces units, for example, have extremely high ammunition consumption rates during training, the shooting conducted within environments made as realistic as possible. By repeating the actions of bringing a firearm to bear on a human target (or at least on a human-shaped target), and doing it constantly, the motion becomes ingrained into the deepest levels of intellectual and physiological memory, to the extent that the actions are performed almost without conscious thought when required. This means that the well-trained soldier is faster to act and respond than the soldier who has received lesser training, a fact that elite military units exploit to the fullest.

So, a key element to controlling aggression is high-quality training, which channels adrenaline into

purposeful action. This principle is applicable to civilian life as well. If there are challenging situations we must face—an intimidating "enemy"—we should train or practice intensively to meet the threat. For example, if you are preparing for a pressurized work meeting about a certain topic, practice your answers to likely questions beforehand, but do it repeatedly until the answers flow out with confidence and assurance. Physically enact the meeting in a place where you can do it unobserved, focusing on making your body language confident, your voice tone controlled and your thoughts clear. (Here is where the acting principle outlined in Chapter 1 comes into play again.)

In conflict zones, soldiers must be ultra-observant for body movements that indicate possible threats. This man could be going for a phone or a gun.

The repetition element here is critical—do the actions until they become totally familiar, but also keep changing the scenario slightly so that you also become accustomed to adapting to unexpected circumstances. Don't just rely upon intellectual preparation. Information is critical and useful, but you must also package its delivery convincingly, which is where physical practice comes into play. In the famous text *Kill or Get Killed*, written in 1943 by American Rex Applegate, an officer in the Office of Strategic Services (OSS), and later adopted as an official U.S. military manual (FMFRP 12-80), the author acknowledges that in the world of unarmed combat, mere book learning is never sufficient:

> *No text, no matter how well illustrated or clearly explained can, alone, teach a man to fight. It can only serve as an instructional guide. Closely supervised intensive practice is the only path to practical knowledge. There are no easy methods or shortcuts. Practice must be intensive enough to render the mechanics of each technique automatic. There is seldom time to stop and think when the pressure of combat is on. Being able to throw a man is much different from knowing how.*
>
> —U.S. Army, FMFRP 12–80, *Kill or Get Killed*, p.4

The key sentence here is "Practice must be intensive enough to render the mechanics of each technique automatic." Although this might appear to apply purely to physical activities, it can also be utilized in the context of mental or social activities too. Essentially, what you are trying to do with such training is turn a desired behavior into a habitual response—something that comes automatically to you when required.

Facing the Enemy

Reality is unforgiving in a combat environment. Any mistakes or miscalculations can be exposed in the most brutal fashion—through fatalities and wounds. Therefore one of the key characteristics of well-trained soldiers is an ability to see the world and the enemy honestly, and to design their actions accordingly.

Above all, soldiers must never underestimate their enemies. There can be a tendency among untrained soldiers to regard the enemy as a static target, sitting passively and therefore exposed to the soldiers' maneuvers and firepower. Good training and operational experience quickly undercuts such notions. The enemy is a calculating and responsive entity capable of taking its own tactical initiative, actively trying to outwit its own enemy (you). The clash between two opponents, therefore, is that of two intelligent groups constantly and fluidly trying to outwit each other.

Here is the essence of unit tactics. Combat leaders must constantly try to maneuver their units into advantageous offensive or defensive positions even as the enemy is trying to do the same, then apply maximum force at the enemy's weakest points to overwhelm and destroy them. In the process, the leaders and soldiers must attempt to put themselves into the mindset of the enemy, attempting to figure out what they would do in the enemy's position.

This is not as straightforward as it sounds. An enemy is likely from a very different cultural background, a background that can soak into and influence their tactical decision-making, even though those tactics are meant to be neutral.

During fighting in the Pacific Theater of World War II, for example, the Americans had to adjust their thinking to the fact that one of the traditional key objectives of military tactics—to stay alive—did not feature very highly among Japanese military forces.

For example, of more than 4800 Japanese soldiers on the tiny atoll of Tarawa, assaulted by the U.S. Marines in November 1943, only 146 were captured, and of those, 129 were actually construction laborers. The rest simply fought to the death, sometimes with suicidal tactics that baffled and terrified the American forces. Consequently, the American tactics came to focus on gaining massive firepower superiority and getting in close to Japanese positions with flamethrowers and explosives, wiping out the enemy with bomb and bullet rather than seeking to force him into a moral collapse.

Rex Applegate put the issue of comprehending the enemy in straightforward terms:

> *The American soldier who meets such an enemy is forced to adapt himself to a pattern of behavior that is foreign to his education and his religious beliefs. If he would win the fight—indeed, if*

he himself would survive—he must know all the dirty tricks of close combat, even as the enemy knows them. He must match them trick for trick. Further, he must be able to take the initiative and attack an enemy soldier as ruthlessly as he, in turn, would be

Left: A fire team—each soldier carries a different form of weapon, to provide a flexible firepower response to threats across a variety of ranges.

Below: The division between combatant and civilian is frequently blurred in modern war zones. Soldiers have to balance friendliness and wariness in equal measures.

"Overwatch" involves providing security and surveillance over other units; here a sniper gives overwatch to an EOD team, which is clearing mines.

attacked if he waited. It is a split second business. There is no time allowed for moral debate.

In close combat, it is now or never. The same principles hold when the enemy is domestic—when he is a brutal criminal running at large; or when he, with other subversives, in a critical hour strikes at our communities. In any case—enemy soldier, dangerous criminal, fifth columnist—the opponent is playing for keeps. Whether we like

> *it or not, we can defeat him and defend our decent standards only by beating him at his own game.*
>
> —U.S. Army, FMFRP 12-80, *Kill or Get Killed*, pp.vii–viii

Applegate urges a ruthless attitude to dominating the enemy, never underestimating that opponent but seeking to crush him when opportunities to do so are identified. So what are the key mental tools required to the take the upper hand?

Coping with Combat Stress

Combat is a true adrenaline experience, alternating (according to soldier accounts) between elation, terror, horror, confusion and satisfaction, depending on the outcome and development of the action. It is undoubtedly stressful, and the body has to make profound adjustments to this situation, as the field manual *Survival* explains:

> *In response to a stressor, the body prepares either to "fight or flee." This preparation involves an internal SOS sent throughout the body. As the body responds to this SOS, the following actions take place:*
>
> - *The body releases stored fuels (sugar and fats) to provide quick energy.*
> - *Breathing rate increases to supply more oxygen to the blood.*
> - *Muscle tension increases to prepare for action.*
> - *Blood clotting mechanisms are activated to reduce bleeding from cuts.*

Combat judgment is based on interpretation. Are these vehicles simply abandoned, or are they part of a road block to set up an ambush?

- *Senses become more acute (hearing becomes more sensitive, pupils dilate, smell becomes sharper) so that you are more aware of your surroundings.*
- *Heart rate and blood pressure rise to provide more blood to the muscles.*

—U.S. Army, FM 3-05.70, *Survival*, 2-6

In combat, these physical responses actually serve a useful purpose, giving the soldier more energy, dynamism and resilience. Yet they are also wearing, and can produce a deterioration in mental acuity, resulting in poor tactical and survivability decisions. So, like aggression, stress must be controlled. There are several methods of doing so, and these methods have equal application in stressful situations in civilian life. The reader should note that these are based upon real-life accounts of soldiers in action, as well as from direct military advice:

Unit tactics are designed to develop all-around awareness. Note how each soldier here has his own designated sector for surveillance and response.

Five key elements of operational discipline: scent, fire, sound, garbage, and light can all give away a position to an observant or proximate enemy.

- **Conscious narration**—either out loud or internally, narrate the situation you are facing in clear and logical language, focusing on using short sentences. For example, a soldier might say amid the hell of a firefight: "We are pinned down in this position. We are taking casualties and the cover is being eroded. We must therefore move position, and do so under covering fire. So, I must define a group to provide the covering fire and identify a position to move to." By expressing thoughts out loud, and in a rather computer-like manner, the soldier can more rationally assess his predicament and make clearer decisions based on facts rather than pure emotion.
- **Breathe**—under stress, breathing patterns can be seriously disrupted, depleting the oxygen levels in the soldier's bloodstream, which in turn clouds judgment. The soldier must therefore remember to breathe, taking long, deep breaths from behind positions of cover to refuel with oxygen and also to steady his thinking.
- **Bias for action**—stress can either paralyze a person with fear or energize him to perform incredible actions. To lean toward the latter, a soldier should be immersed in positive action rather than a despairing passivity. Amid the confusion, the soldier should keep

his focus upon the objectives at hand, and seek out routes of action that take him toward fulfilling these objectives. With a point of concentration, the soldier is better able to channel his stress into action rather than let it envelop him.

Leadership

Leadership, as already noted, is a key determiner of the success of a military unit, particularly in combat. The officer, and the NCOs beneath

A simple representation of the flanking tactic. Tactics are conceived in clarity, but executed in a far messier reality.

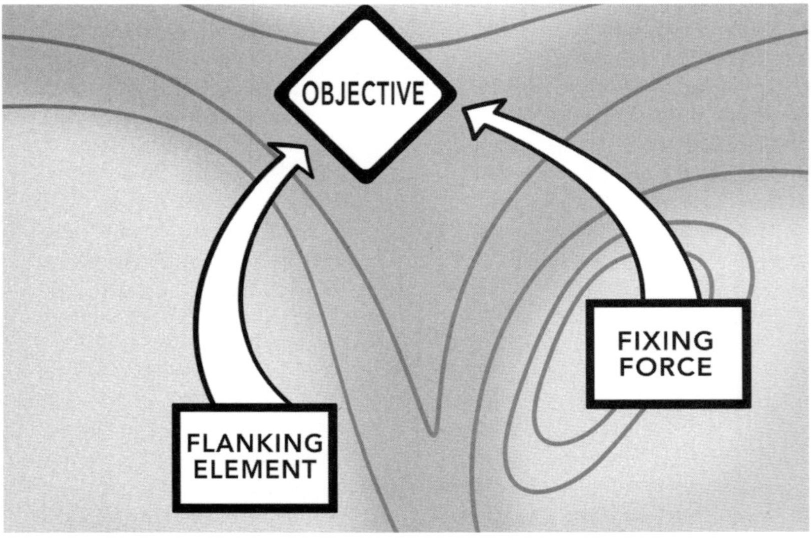

him, should project confidence, clean decision-making and a general unflappability at all times. All fear and anxiety is contagious in a military unit, but the spread of this disease is particularly virulent if it starts visibly in an officer.

There are certain identifiable traits that seem to be characteristic of effective tactical decision-making in leaders, and we will touch on some of the more important ones here. Note, however, that although these are leadership properties, they are not confined to officers per se, but to any soldier who needs to take charge against an obdurate enemy. In effect, they are qualities of self-leadership

During battle, soldiers must constantly think about how they can position themselves behind cover, and move quickly between pieces of cover.

as much as team leadership. These are derived from the Marine Corps *Warfighting* book mentioned previously, which is required reading for all those wanting to understand the military mindset.

Speed and Focus

Speed is a relative advantageous rapidity of movement against the enemy, maneuvering at a velocity that the enemy cannot match in his

Leadership Development

The following advice is given to enable an individual to perform an honest evaluation of his or her leadership qualities, and how they can be improved:

Complete a Situation Analysis for Self-development
Think of experiences over the past two years that give insight into personal strengths and developmental needs—maybe a critical decision, an important task you led or were a part of, or a significant personal interaction. Use these questions to analyze each situation:

- What was the situation? What was happening? Who was there?
- What was the goal and did you reach it? What were you trying to accomplish? What resources or skills did you have or not have that you needed?
- What did you say and think? Were you able to find the right words to make your point? What were you thinking at the time? What made you feel good (confident, excited) or bad (confused, worried)?
- What did you do? How did you act (including your body language)? How did others react? Did you help or hurt the situation? Did you adjust your actions based on how others were reacting?
- Why did you act the way you did? What knowledge and skills led you to act the way you did?
- What could have helped you handle the situation better?
- How could you have used your strengths to reach a better outcome?
- Are there any developmental needs that you should make a high priority for personal self-development efforts?

—U.S. Navy, *Naval Swimmer and Instructor's Manual,* 8.5

responses. The German *Blitzkrieg* campaigns of 1940, for example, were so effective because they used armored thrusts and combat aviation to keep the British, Belgian and French opponents wrong-footed and either trapped in useless defensive positions (which the Germans often handled by simply bypassing them) or constantly on the retreat to try to gain a defensive position. Moreover, the speed also had *focus*, which means channeling the movement into defined objectives. Of focus, *Warfighting* states: "We achieve focus through cooperation toward the accomplishment of the common purpose. This applies to all elements of the force, and involves the coordination of ground combat, aviation and combat service support elements." So focus is at its heart teamwork; good combat leaders should seek to work with all parties involved, not put the whole enterprise at risk through glory-seeking unilateral behavior.

The virtues of speed are also accentuated by the application of *tempo*. Tempo is *speed over time*, and if maintained tempo results in an enemy who is constantly unable to catch their breath and recover the initiative. Note that maximum speed is impossible to maintain indefinitely, but tempo at a reasonable speed

In this diagram of a pincer-movement attack, the enemy is "fixed" in place by a frontal assault while two other elements attack the more vulnerable flanks.

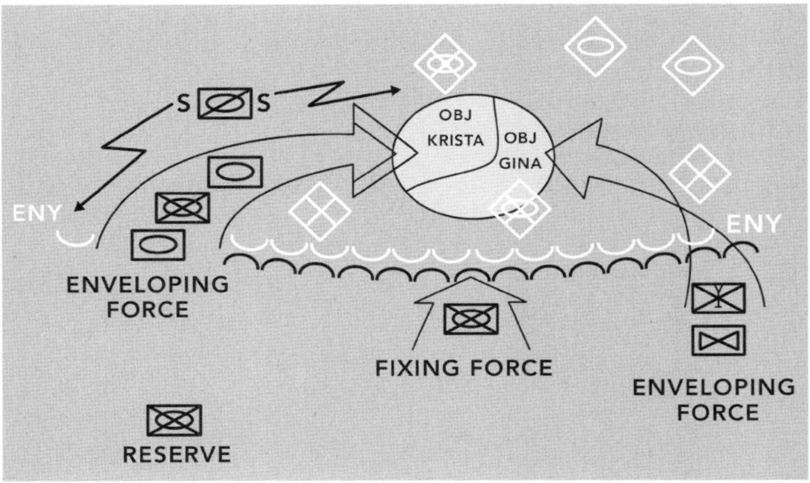

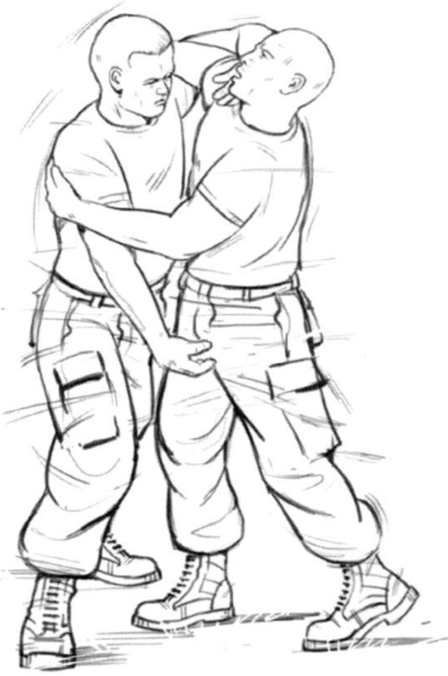

Military-style unarmed combat is unrestrained by any sort of social rules. Each attack is aimed to debilitate, injure or kill the opponent.

can be prolonged, giving the overall advantage on the battlefield. The civilian lessons here are clear. If you are in a competitive situation, act fast and with clear purpose, and maintain the tempo to overwhelm and exhaust your opponent.

Surprise and Boldness

Surprise is "a state of disorientation resulting from an unexpected event that degrades the enemy's ability to resist" (*Warfighting*, p.42). It is achieved by three main methods—deception, ambiguity (maneuvering in a way that the enemy cannot interpret) and stealth—and critically it is about forcing a confused reaction from the enemy and taking advantage of that reaction. Surprise can be difficult to achieve, especially as the enemy will, if they know what is good for them, be monitoring what their opponents are up to. Yet it is achievable on both small and great levels.

A former U.S. Marine officer, a Vietnam veteran, once told me about surprising a group of Viet Cong (VC) insurgents with an ambush on a trail in the South Vietnamese jungle. In a grim counterpoint to the ambush, the officer had predicted that the VC soldiers, once attacked, would

A basic fighting stance, with the hands at guard to protect the torso and jawline, the feet positioned for quick movement.

Above: Soldiers can be faced with the need to make rapid judgment calls. Is the approaching vehicle a confused civilian or a suicide bomber making an attack?

leap for cover into a ditch on the far side of the trail. Thus before the VC marched into the area, the officer had laid det-cord (explosive cord used for connecting demolition charges) in the bottom of the ditch. As predicted, the VC men leapt into the trench to avoid the ripples of sudden small-arms fire, only to be blown up by the det-cord.

On a larger scale, the Allied forces in Europe during World War II used deception, ambiguity and stealth on an epic scale to hide the fact that Normandy was the intended location for the opening of the second front in

World War II, the operation launched in June 1944.

Boldness goes hand-in-hand with surprise: "Boldness is the characteristic of unhesitatingly exploiting the natural uncertainty of war to pursue major results rather than marginal ones... Boldness is superior to timidity in every instance although boldness does not always

Below: Good leaders are capable of seeing through the "fog of war" to issue clear and focused commands leading to purposeful actions.

equate to immediate aggressive action. A nervy, calculating patience that allows the enemy to commit himself irrevocably before we strike him can also be a form of boldness. Boldness is based on strong situational awareness: we weigh the situation, then act. In other words, boldness must be tempered with judgment lest it border on recklessness" (*Warfighting*, p.44).

Boldness and surprise are characteristics that can be applied in civilian life with equally impressive effect. The history of entrepreneurship provides clear evidence that most of the really

Combat can provoke panic reactions in some people. While the rest of the unit has a duty of care, in combat they must keep focus on the mission.

impressive gains are by those who ignore, or at least aren't constrained by, convention, and who apply themselves to projects that are daring and provocative. Of course, many of these ambitions have led to failures, but others have led to the greatest company successes in history.

Centers of Gravity and Critical Vulnerabilities

"Centers of gravity" is a useful concept for determining the weaknesses of an enemy. In military thinking, it refers to the fact that an enemy is not a homogenous entity, but rather a mixture of physical, moral and social relationships, with some

Professional soldiers have a keen sense of awareness. What appears to be a sixth sense for trouble is often gained through hard-won experience.

Initiative, Judgment, and Teamwork

Initiative and judgment are essential in both peacetime and combat. On the battlefield soldiers need initiative to operate within the intent of their commander and to move decisively in accomplishing their mission. This is true whether "combat action" involves firing at the enemy, performing maintenance and repairs on combat equipment or driving a truck that takes essential food, ammunition or fuel to the battle. However, initiative does not mean "do something even if it's wrong."

Initiative must be tempered by good judgment—the ability to size up a situation quickly and to know what is important and how to accomplish what needs to be done. Soldiers with initiative tempered by good judgment act on their assessments quickly and decisively with little or no supervision. They accept responsibility and take thoughtful action to operate successfully and to execute difficult missions.

Another ingredient of successful teamwork is confidence. To remove doubt and anxiety in combat, the soldier must first have confidence in his own professional ability. Then he must be confident that his fellow

of the aspects more important than others for the integrity of the whole. *Warfighting* clarifies the proposition: "Which factors are critical to the enemy? Which can the enemy not do without? Which, if eliminated, will bend him most quickly to our will? These are centers of gravity."

For example, an army's communications systems might be a center of gravity. Saddam Hussein's forces discovered this reality during the 1990–91 Gulf War, when Allied aircraft progressively wiped out the Iraqi command-and-control systems, effectively rendering the armed forces "blind" and unable to respond in a

An officer uses a sand table to brief a mission, the miniature landscape representing the area of operation.

team members, as well as other supporting soldiers, can do their jobs effectively. For example, when a forward observer calls for "danger close" fire, the soldier needs confidence in the accuracy of the forward observer and of those delivering the fire. The soldier needs to feel confidence in his leader. The leader earns his soldiers' confidence as he demonstrates his ability to do his job. Soldiers and leaders develop mutual confidence by sharing difficult, challenging and realistic training, as well as the rigors and dangers of combat. Mutual confidence multiplies combat power as it welds individuals into cohesive teams.

—U.S. Army, FM 22-100, *Military Leadership*

A fire team moves forward under mortar fire, keeping its formation despite the threat and the danger.

coordinated fashion to the coalition maneuvers. Military snipers will devote much of their attention to identifying and killing enemy officers, because leadership is typically a center of gravity in any military formation.

In urban warfare, the key tactical problem is that the landscape is fractured into multiple individual buildings and complex lines of movement.

Critical Vulnerabilities

Allied to the center of gravity proposition is that of "critical vulnerabilities." These are the areas in which the enemy is weakest and most vulnerable to attack. They might be physical vulnerabilities, such as an opening in a defensive line, or they might be human vulnerabilities, such as dependence upon a particular leader or problems obtaining adequate food supplies. From a tactical point of view, it is obviously better to direct your strengths against an enemy's weak points to maximize the effect of the punch.

Note that centers of gravity and critical vulnerabilities are essentially complementary. The former focuses on the enemy sources of strength, and the latter on the enemy sources of weakness, and attempts to turn both to the friendly forces advantage. From a civilian perspective, both concepts are useful because they help us to think of ways to tackle a problem without necessarily going at it head on.

As an example from my own history, when I was at university I naturally looked for a way to maximize the grades of my papers.

A Spetsnaz soldier of the Soviet era showing some of his extensive kit and weaponry, including a Dragunov rifle.

One thing I realized, through observation, was that when a paper was assigned, the university library was most crowded in the sections that held all the references for that particular subject. In other words, the vast majority of students were researching from the most common and recommended sources. I reasoned that this must produce a large volume of papers with a similar approach to the essay question, which in turn must be quite boring for the teaching assistant (TA) to grade. (Having performed the grading role later in my academic life, I can confirm that this is true.) So I reasoned that boredom was in essence the grader's critical vulnerability, and his center of gravity was the need to see something interesting and original.

To focus on these factors, I therefore did my paper research using areas of the library with a tangential connection to the essay topic, forcing me to make original links between

Boxing is used by several elite military units to evaluate the soldiers' ability to control and unleash their aggression.

different disciplines. Doing so, the TAs later told me, meant that some of my essays stood out from the pile, resulting in higher grades. Note that this approach had nothing to do with intelligence, but everything to do with method.

Dominating an enemy never has predictable outcomes. What this chapter has shown, however, is that the intelligence and mental attitude you bring to a fight, plus the quality of the training, can be the decisive factors in securing a victory. Such is even the case for the most complex enemy of them all—terrorists—the focus of our next chapter.

U.S. soldiers fire their M4 Carbines. Modern combat optics have made the assault rifle a precision tool.

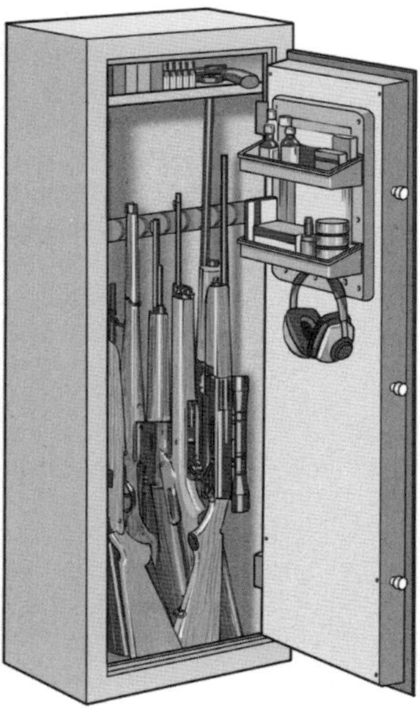

4

Coping with Terrorist Attacks

Modern terrorism is as much about perception as reality. While the world has to confront terrorist actions, it also has to battle the fear that is actually terrorism's strongest weapon.

Terrorism is the shadow under which virtually all societies now live. From a complex spectrum of political, ideological, cultural and historical causes, modern terrorism emerged in earnest during the 1960s and 70s and has remained with us ever since. The nature of terrorist attacks has changed over time. During the 1960s to 1980s, the major focuses of the terrorists were prominent organizations, buildings, military forces and individuals, the targets selected for their perceived impact upon government thinking and corporate wealth. This is why aircraft hijacking was so popular, as the generally international composition of aircraft passengers enabled the terrorists to apply pressure on multiple governments simultaneously.

Yet from the 1990s onward, terrorism shifted its gaze somewhat away from "hard" targets such as airliners or government buildings, not least because the rise of specialist counterterrorism forces and anti-terrorist countermeasures made such targets more difficult to strike. Instead, terrorism reached out to hit civilian populations, innocent people no longer classed as innocent when viewed through the ideological lens of fundamentalist religion—the dominant source of

Opposite: Terrorism today is a threat to all societies.

terrorism or political aggression—in our modern age.

The outcome has been a series of horrific attacks that have, in turn, led to major regional conflicts. They include the stunning aerial attacks on the World Trade Center and the Pentagon in 2001 (which directly produced the subsequent wars in Afghanistan and Iraq); the train bombings in Madrid in March 2004; the sequence of deadly bombings in London on July 7, 2005; the four-day massacre in Mumbai in 2008; and the recent Paris attacks of November 2015 and Brussels in March 2016. These attacks not only resulted collectively in thousands of people dead and wounded, but have led to the increasing sensation that nowhere is truly safe from terrorism.

At a training camp, would-be terrorists practice their small-arms skills. Terrorist training has become increasingly professional.

Against this backdrop, however, we must remind ourselves that passivity against terrorism is not an option. Indeed, we must not overstate the strength of terrorists, who are killed or captured regularly by state forces, and live out much of their short lives in clandestine anxiety. In this chapter, we will look at how mental attitude and applied intelligence can help us all in the fight against terrorism, ensuring that we are not conquered by our own fear.

Terrorist groups tend to fare badly when confronted by professional military forces, as they often lack the tactical discipline needed to engage competent armies.

The Terrorist Mindset

In the previous chapter, we discussed the importance of understanding your enemy as an active and thinking entity, and one that must not be underestimated. Getting into the mindset of your enemy is particularly crucial in the fight against terrorists. Such organizations are propelled to act by a complex mixture of ideological, religious, cultural, military and social drivers, and a lack of appreciation for these factors will weaken any counter response to terrorism. In 2014, the U.S. government released its new counterinsurgency manual, a large work devoted to explaining the principles of combating terrorism and insurgency for both regular forces and specialist troops alike. The manual makes some emphatic points about understanding the enemy:

> *The way that a culture influences how people view their world is referred to as their worldview. Many people believe they view their world accurately, in a logical, rational, unbiased way. However, people filter what they see and experience according to their beliefs and worldview. Information and experiences that do not match what they believe to be true about the world are frequently rejected or distorted to fit the way they believe the world should work. More than any other factor, culture informs and influences that worldview. In*

Classic terrorist tools: A suicide vest, an AK-47, a handgun, a grenade, IEDs (improvised explosive devices—here two artillery shells) and, for communications, the humble cellular phone.

A modern U.S. Psychological Operations (PsyOps) vehicle, fitted with a top-mounted loudspeaker to broadcast messages to local populations.

other words, culture influences perceptions, understandings and interpretations of events. Soldiers and Marines need to know that U.S. interpretations of events are often quite different from the perceptions of these events by other people in an area of operations. If soldiers and Marines assume that the local population will perceive actions the way that they do, they are likely to misjudge their reactions. The U.S. military refers to this pattern of assuming others see events in the same way the U.S. does as mirror imaging. Mirror imaging is dangerous because it leads soldiers and Marines into thinking that their assumptions about a problem and its solution are shared by the population and multinational partners, rather than employing perspective taking, and looking at the problem from the population's perspective.

—U.S. Army, FM 3-24/MCWP 3-33.5, *Insurgencies and Countering Insurgencies*, 3-1

The point this passage makes about "mirror imaging" is crucial to the counterterrorism response. Only by understanding the genuine differences between their own culture and the foreign culture can soldiers attempt to deliver the "hearts and minds" operations that are actually the best countermeasure against terrorism. By building sympathetic relations with a host culture, military forces are better able to deprive terrorists of local support, in the process isolating them and hopefully making them more visible for forcible action.

In actuality, the cultural battleground is far more important than the tactical battleground when it comes to fighting terrorism. Today, professional armies invest considerable time and money training their soldiers to understand host cultures and to try to build bridges between the soldier's world and other people's. The investment is designed to develop what is known as "cultural and situational understanding." This type of understanding is only partly about the knowledge soldiers have of the surrounding cultural reality. They

Terrorist leaders value targets that a) maximize the potential for inflicting casualties; and b) carry some form of political symbolism.

certainly must be aware of the ethical, social, gender and historical context of their theater of operations, but they must also be able to empathize with the local people and their situation, and be able to comprehend an apparently alien culture in terms of common human drives.

For however different a people or country appears to be when compared to our homeland, there is always as much uniting us as separating us. Shared human instincts include the need for dependable sources of food and shelter, the basic desire for education and for employment, safety for family and friends, and the need to be respected and valued. By understanding both the shared and the different experiences between cultures, a soldier is better placed to sense when something feels right or suspicious, and to react accordingly.

An IED device emplaced inside a small suitcase. Remote explosive attacks are the most influential tool in the terrorist arsenal.

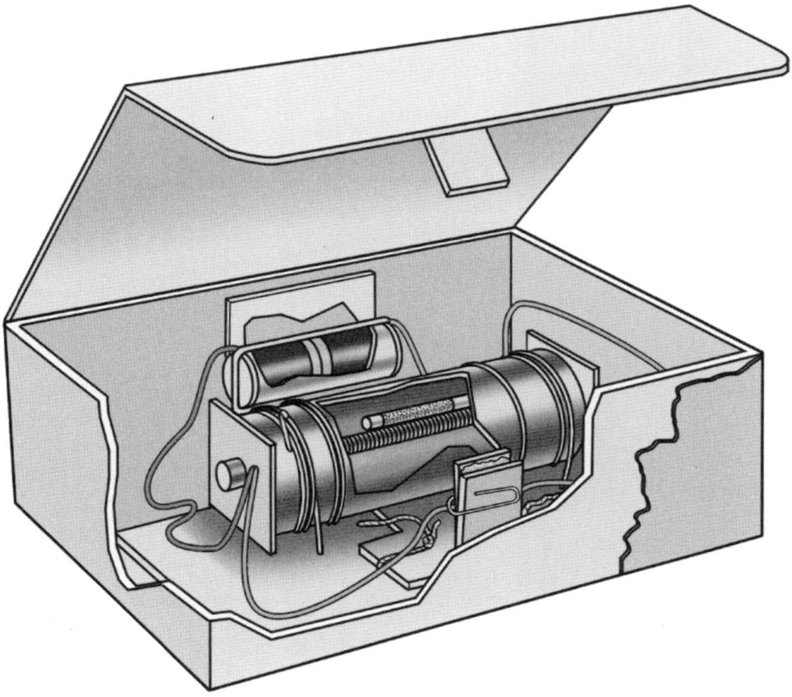

Understanding a foreign culture fully can be something that takes a lifetime to achieve, depending on the sensitivity of the individual. Yet there are several key areas of understanding that should take priority in the soldier's thought processes and research if he or she is to avoid causing offense to local people, and thereby act as an unwitting recruiter for the enemy.

Social organization—how is the society organized, particularly at the family, tribal and regional levels? Who holds authority in that society, and what is the relationship between family/kinship groups and official governmental authorities? Understanding a culture's social organization provides for more focused and effective negotiations with local people.

Family and gender relations—family and gender relations can be extremely different from country to country. In the Middle East and Asia, for example, the strength and integrity of the extended family (especially relations such as cousins and uncles) can be far more important than it is in many Western cultures. Within the family, and in society at large, gender relations can also be more rigidly hierarchical, with a higher degree of male authority over females, and far stricter codes regarding sexual behavior and styles of dress. Understanding both family structures and gender relations is crucial if a

Soldiers must be capable of restraint during riot-control duties, as excessive force can simply breed more recruits for the terrorist cause.

soldier is to avoid causing offense. For example, what might be seen as harmless flirting in Western Europe or the United States could be seen as a shocking insult to religious codes throughout much of the Middle East.

Religion—faith is the cultural foundation of many world societies, with religious concepts holding influence at the very highest levels of government and social organization. A soldier should therefore seek not only to understand the key tenets of the religion, but also how the religion is expressed ritualistically and practically in society, including in aspects such as law and business

Terrorism is frequently born out of existing civil conflict. Soldiers in counterterrorism should invest in researching the historical background of the situation.

transactions. Critically, the soldier must understand the codes of appropriate behavior (including dress and language) around religious sites and buildings, plus note in their diaries any special religious festivals or holy days, which can be markers for increased terrorist activity.

History—some historical understanding of the host culture is essential to knowing why the people

think and behave in the way they do. For example, in many countries with histories of dictatorships or totalitarian rule, asking detailed questions to solve problems is often avoided, at least with people whom the questioner does not trust, because asking questions exposes your ignorance to the authorities and in turn also raises questions about why you are looking for the answers. Or the population might appear generally noncompliant to the foreign military force, even if that force is apparently benign, because the population has experience of previous armies moving through and then leaving, having not resolved the problems they said they would fix. Soldiers should also explore specific local histories, such as the

For soldiers, counterterrorism can be a uniquely stressful form of warfare. Counseling should be sought if the duties start to inflict serious psychological symptoms.

relationships between powerful tribes and families. Many societies operate upon an honor code, meaning that revenge has to be enacted to satisfy a slight or violence from another tribe or family. The honor code can result in literally decades of conflict at a local level, rumbling on independently of larger national wars but often intersecting with those conflicts. In Afghanistan, for example, it has often been found that a determiner of whether a group belongs to the Taliban or is allied to the Western forces depends on how the chosen alliance provides an advantage against another tribe or family grouping.

Appropriate behaviors—this is a blanket term for a bewildering array of customs, gestures and interactions, some of which can be totally unique to a single country. A soldier from a foreign country is unlikely to understand all such behavioral nuances, but he should at least become familiar with the following:

- Customary greetings.
- The appropriate treatment of women and children.
- Common courtesies (such as gift giving).
- Accepted formulas for conducting negotiations.
- Understanding of the notion of time (for example, some cultures do not have the same sense of punctuality and urgency as is common in the West).
- Acceptable and unacceptable hand gestures.
- The temporal structure of daily life, such as mealtimes and working hours.
- How to open conversation appropriately and what topics are acceptable for discussion with strangers.

Language—of course, being able to understand and speak the local language gives the soldier the greatest insight into the local

A terrorist nuclear-biological-chemical (NBC) attack is a nightmare scenario. Here citizens wear face masks to protect against airborne threats.

culture. Language is not just about comprehending what is being said, but also contains the fundamental conceptions of how a culture is constructed and how it views the world. Even with just a basic grasp of the local language, the soldier will be able to break down barriers between himself and those around him, and also be more capable of picking up valuable intelligence. Most military organizations now provide access to language courses, and it is recommended that any soldier who is going on foreign deployment should attend a relevant course.

By mastering the basics of the foreign culture, a soldier will be far more "primed" to spot anomalies in that culture, and therefore to detect terrorist activity. In Afghanistan and Iraq, for example, a classic sign of an impending attack is the absence of children playing in the streets—their families have been pre-warned to keep their children inside so they don't get caught up in the violence.

Cyber warfare is now at the frontline of the war on terrorism, especially in terms of data security and terrorist surveillance.

Communications is one area in which modern armies can establish a clear advantage over terrorist groups; here a soldier monitors insurgent "chatter."

The mix of military and civilian dress here could mean that this individual has allegiances to a paramilitary group.

Collecting Intelligence

Having become immersed in the local culture, the soldier will then attempt to acquire as much counterterrorist intelligence as possible. While electronic intelligence gathering is known as signals intelligence (SIGINT), the extraction of intelligence from human sources is known as human intelligence (HUMINT). HUMINT is one of the greatest weapons against terrorist groups. Good-quality HUMINT can provide everything from the location of a terrorist leader to the modus operandi of a terrorist attack. The key to collecting HUMINT is alertness, being constantly observant about every detail of the surroundings—nothing is irrelevant—but also being able to process the information gathered and convert it into usable plans of action. Again, the U.S. military has produced a highly informative manual on this subject entitled *Human Intelligence Collector Operations* (FM 2-22.3). It is an exhaustive document, explaining in detail personal qualities required together with good intelligence. The following passage explains in detail just some of the personal qualities needed to be a capable HUMINT gatherer:

> ***Alertness***
>
> *The HUMINT collector must be alert on several levels while conducting HUMINT collection. He must concentrate on the information being provided by the source and be constantly evaluating the information for both value and veracity based on collection requirements, current intelligence and other information obtained from the source. Simultaneously, he must be alert not only to what the source says but also to how it is said and the accompanying body language to assess the source's truthfulness, degree of cooperation and current mood. He needs to know when to give the source a break and when to press the source harder. In addition, the HUMINT collector constantly must be alert to his environment to ensure his personal security and that of his source.*
>
> ***Objectivity and Self-control***
>
> *The HUMINT collector must also be totally objective in evaluating the information obtained. The HUMINT collector must maintain an objective and dispassionate attitude regardless of the emotional reactions he may actually experience or simulate during a questioning session. Without objectivity, he may unconsciously distort the information acquired. He may also be unable to vary his questioning and approach techniques effectively. He must*

have exceptional self-control to avoid displays of genuine anger, irritation, sympathy or weariness that may cause him to lose the initiative during questioning but be able to fake any of these emotions as necessary. He must not become emotionally involved with the source.

Adaptability

A HUMINT collector must adapt to the many and varied personalities which he will encounter. He must also adapt to all types of locations, operational tempos and operational environments. He should try to imagine himself in the source's position. By being adaptable, he can smoothly shift his questioning and approach techniques according to the operational environment and the personality of the source.

Perseverance

A tenacity of purpose can be the difference between a HUMINT collector who is merely good and one who is superior. A HUMINT collector who becomes easily discouraged by opposition, non-cooperation or other difficulties will not aggressively pursue the objective to a successful

Police searches often become routine during times of terrorist threat. Such searches hamper terrorist freedom of movement.

conclusion or exploit leads to other valuable information.

Appearance and Demeanor

The HUMINT collector's personal appearance may greatly influence the conduct of any HUMINT collection operation and attitude of the source toward the HUMINT collector. Usually an organized and professional appearance will favorably influence the source. If the HUMINT collector's manner reflects fairness, strength and efficiency, the source may prove more cooperative and more receptive to questioning.

Initiative

Achieving and maintaining the initiative are essential to a successful questioning session just as the offensive is the key to success in combat operations. The HUMINT collector must grasp the initiative and maintain it throughout all questioning phases. This does not mean he has to dominate the source physically; rather, it means that the HUMINT collector knows his requirements and continues to direct the collection toward those requirements.

—U.S. Army, FM 2-22.3, *Human Intelligence Collector Operations*, 1-11–1-12

When large numbers of prisoners are taken, it becomes imperative to separate them as quickly as possible, to prevent the development of cover stories.

The manual goes on to list a whole spectrum of other traits needed for the collection of HUMINT. The type of person required is tenacious, sensitive to the surroundings, imaginative but realistic and does not impose his own preconceptions on the information. In terms of objectives, the following are key informational goals for the HUMINT operator fighting terrorism:

- Known members of terrorist organizations (particularly leaders), and where they live/ operate.
- Training bases used by terrorist groups.
- Any information leading to the identification of foreign support networks.
- Planned terrorist operations.
- Methods by which the terrorists recruit new members.
- Sources of finance, supplies and weaponry for the terrorist group.
- Any factions within the terrorist organization.
- The preferred communications system used by the group.
- The level of local support for the terrorist aims.
- Terrorist sources of information regarding the soldier's own armed forces.

Avoiding and Responding to Terrorist Attack

At times or in places that are susceptible to terrorist attack, both soldiers and civilians must exercise high levels of personal and situational awareness. You should avoid predictable routines and journeys, for these are conducive to terrorist attack planning. In public spaces, be vigilant for any behaviors by individuals or vehicles that appear

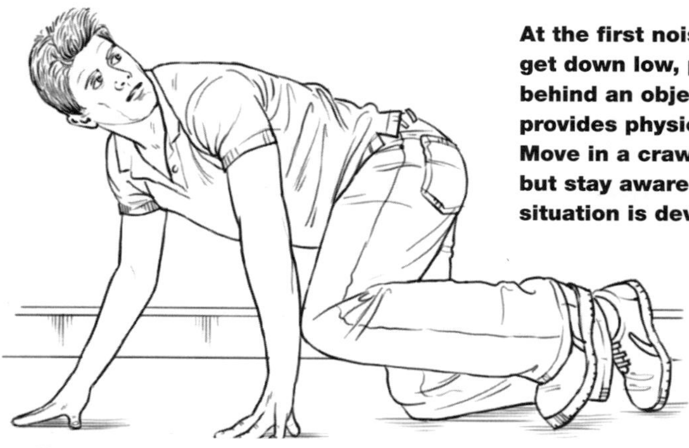

At the first noise of gunfire, get down low, preferably behind an object that provides physical protection. Move in a crawl if necessary, but stay aware of how the situation is developing.

Left: A sturdy table is a good object under which to hide in the case of a bombing, providing some degree of overhead protection.

suspicious and out of place. For example, be alerted by:

- Individuals wearing unseasonal clothing and looking nervous.
- Individuals watching a public venue intently and frequently communicating on a mobile phone.
- Vehicles repeatedly driving past a specific location (although obviously rule out the fact that

Below: Learn how to perform basic resuscitation techniques as part of your security training.

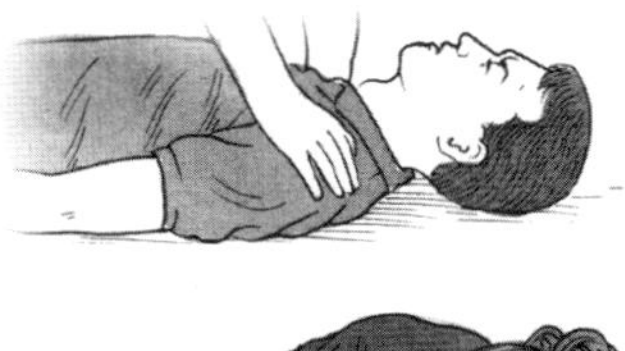

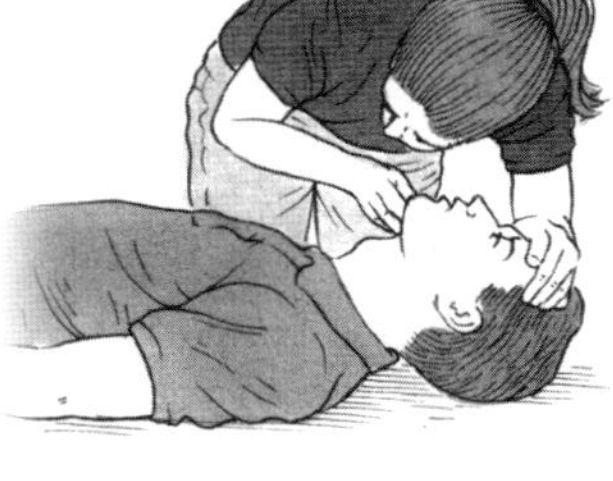

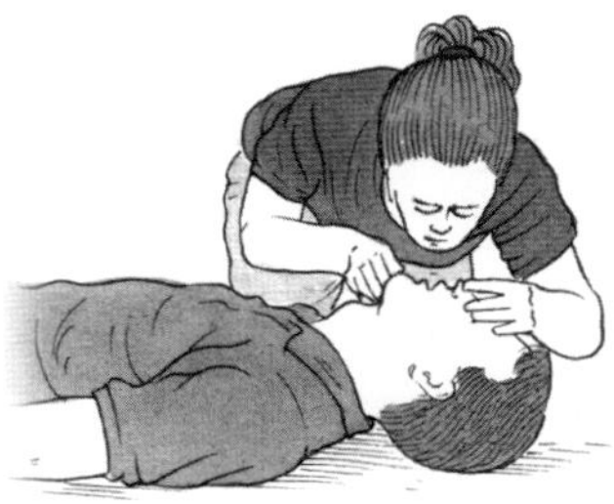

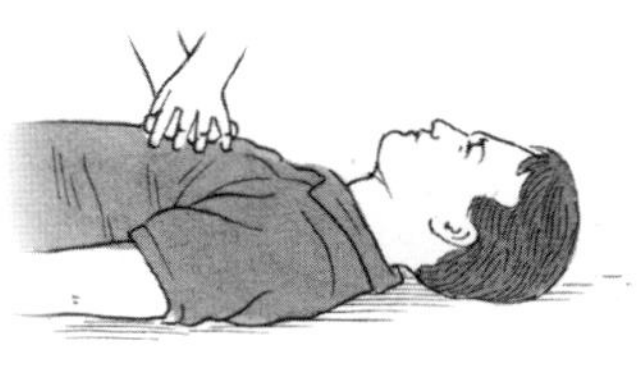

the drivers might be looking for a parking spot).

- Individuals or groups of individuals moving quickly through public spaces, particularly if they are carrying heavy-looking packs or bags.
- Individuals depositing bags in public places and then walking away from them.

To aid your awareness of such developing situations, think about your movement and positioning in public places. Try to gain a good vantage point. On trains, try to stand at one end of the carriage near a door, so you have a good view along the entire length of the carriage plus you have a nearby route for entry and exit. In restaurants, bars, busy offices

Right: Here the first aider has identified the correct position for chest compressions, the heel of the hand going on the sternum.

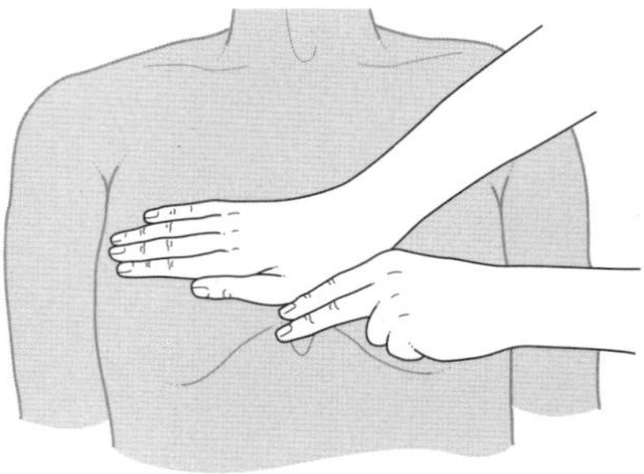

Below: Bleeding can be controlled by placing a thick, clean pad on the injury site and maintaining a firm, prolonged pressure.

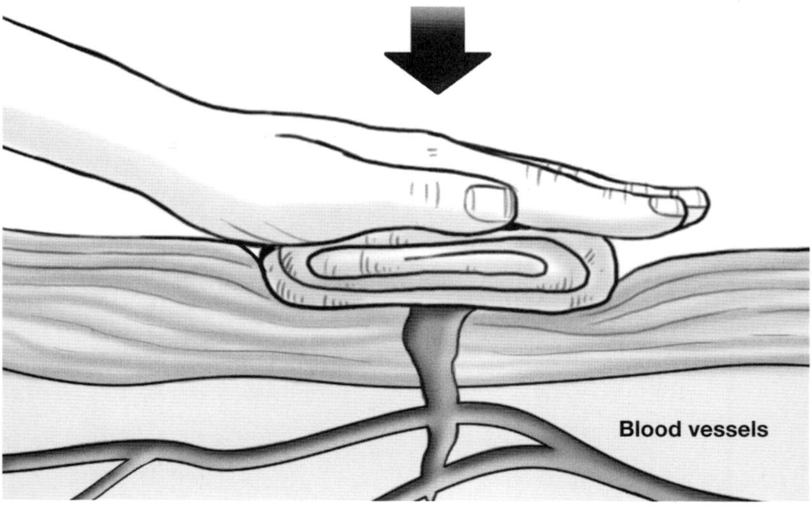

or other places of social gathering, position your body with a solid wall behind you and clear views of all entry points around the space. Ask yourself—if attackers came in through the door, where could I maneuver or

Many people feel the best preparation for a terrorist attack is a well-stocked firearms cabinet; others might argue that the availability of firearms makes the situation even more dangerous.

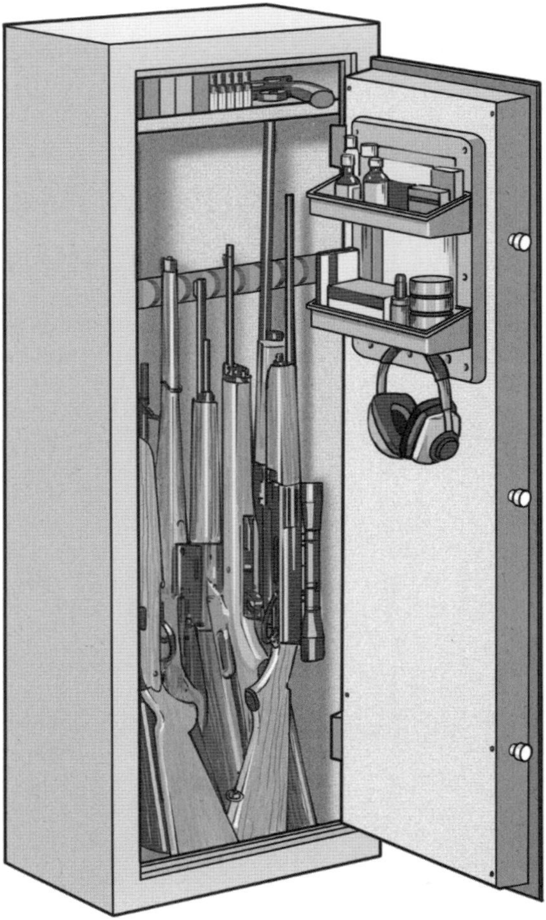

Hard Targets, Soft Targets

The following passage comes from the U.S. Marine Corps manual MCRP 3-02E, *The Individual's Guide for Understanding and Surviving Terrorism*. It explains how government employees and military personnel should adapt their behavior in high-threat environments to reduce the danger from terrorist attack:

Soft targets are accessible, predictable and unaware. They make it easy for strangers to access their private information (e.g., phone numbers, addresses, schedules). Soft targets follow consistent routines at home and at work. This allows terrorists to predict their movements in advance. They are unaware of their surroundings and do not employ individual protective measures.

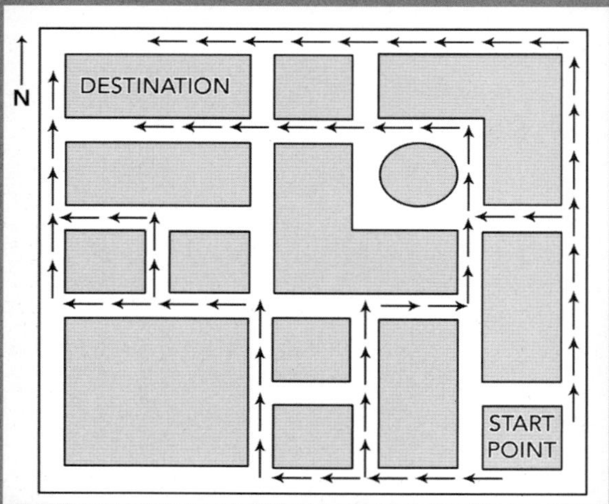

In high-risk areas, vary your routes to and from a frequent destination, to confound terrorist planning and intelligence gathering.

Hard targets are inaccessible, unpredictable and aware. They make it hard for terrorists to gain access to themselves or their families. Hard targets consciously vary their routines and avoid setting patterns in their daily life. They are security conscious, aware of their surroundings and proactively adhere to individual protective measures. Hard targets do not:

- Put their names on mailboxes or exterior walls of their house.
- Run or walk every day at the same time or at the same place.
- Wash cars, mow lawns or have family cookouts the same day every week.
- Shop the same day of each week at the same store.
- Travel to and from home on the same route and at the same time.
- Attend church services at the same time and place each week.
- Sit in the same seat in a vehicle, restaurant, church, etc.
- Take the same route to work every day, arrive at work at the same time, go to lunch at the same time or leave work at the same time each day.
- Pick up the newspaper or mail at the same time every day.
- Walk or feed the dog along the same route or at the same time every day.
- Go to the same restaurants or bars or patronize only American restaurants or bars.
- Park their vehicle in the same location when going to work, church, social events, etc.
- Earn the reputation of always lending a helping hand.

– U.S. Marine Corps, MCRP 3-02E,
The Individual's Guide for Understanding and Surviving Terrorism,
2-1

In Iraq, IEDs were emplaced in every conceivable location, even within dead animal carcasses and plaster of Paris curbstones.

hide? And, of course, never neglect the fundamentals of security—keep doors and windows locked and shut; in public buildings implement proper personnel checking procedures at the entrance; scan around and beneath vehicles before getting into them to check for suspicious devices; steer clear of suspicious packages; avoid areas that you know to be threatening.

The fact remains, however, that there is definite limit to what you can do to protect yourself against terrorist attack, for the terrorists innovate in the offensive as much as you do in the defensive. There are two likely nightmare scenarios for a terrorist incident—a bombing and a co-ordinated shooting. (Hostage taking is covered in the next chapter.)

Your mental response to these events can, depending on the situation, determine your potential survival. What is critical is that you do not go into denial about the attack, but respond promptly to it with decisive action. In the case of a bombing, assuming that you have survived the initial blast unscathed, always bear in mind that there might be a secondary device waiting to go off.

For this reason, unless you're called upon to provide physical help to the wounded, move quickly away from the attack area, creating as much distance as possible between you and any suspiciously parked vehicles—remember that the blast

When moving during a terrorist attack, stay behind cover and keep below the possible line of sight of enemy shooters.

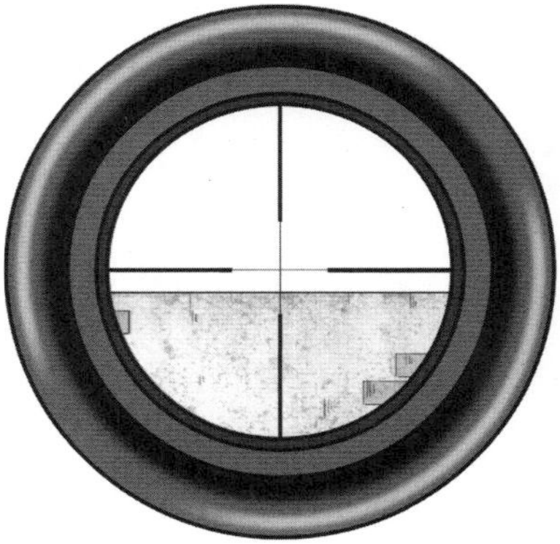

This woman has picked excellent cover during an attack; the solid iron girder would stop all but the heaviest projectile, and it shields her entirely from view.

radius of a bomb can be several hundred yards. If you are inside a building and a device goes off outside, stay away from windows, doors and elevators in case another device is about to detonate.

As the people of Paris recently and tragically experienced on November 13, 2015, one of the most terrifying forms of attack is a coordinated shooting delivered by multiple gunmen in public areas. Such attacks are designed to maximize confusion and horror, channeling large groups of people into killing zones, usually implemented by individuals who know that the operation will end in their own deaths.

In response to the attacks in France, the British government issued its own guidelines about the correct response to terrorist assault of a similar type on British soil. The advice is as follows:

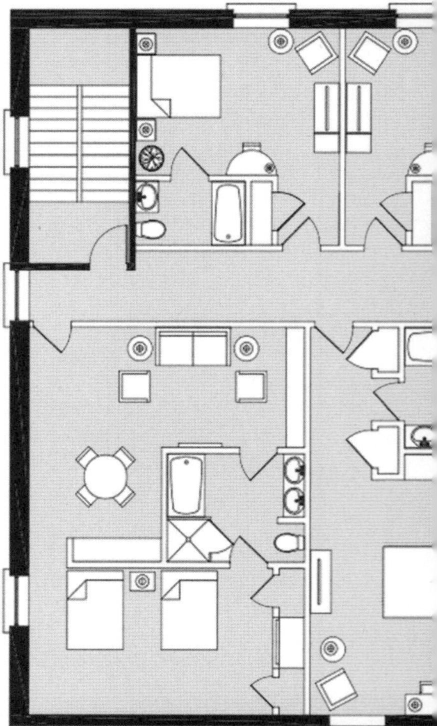

Run

- *Escape if you can.*
- *Consider the safest options.*
- *Is there a safe route? RUN if not HIDE.*
- *Can you get there without exposing yourself to greater danger?*
- *Insist others leave with you.*
- *Leave belongings behind.*

Hide

- *If you can't RUN, HIDE.*
- *Find cover from gunfire.*
- *If you can see the attackers, they may be able to see you.*
- *Cover from view does not mean you are safe, bullets go*

If you are staying in a hotel in a high-risk area, get to know the building plan with all the major entrances and exits, plus access points to service areas.

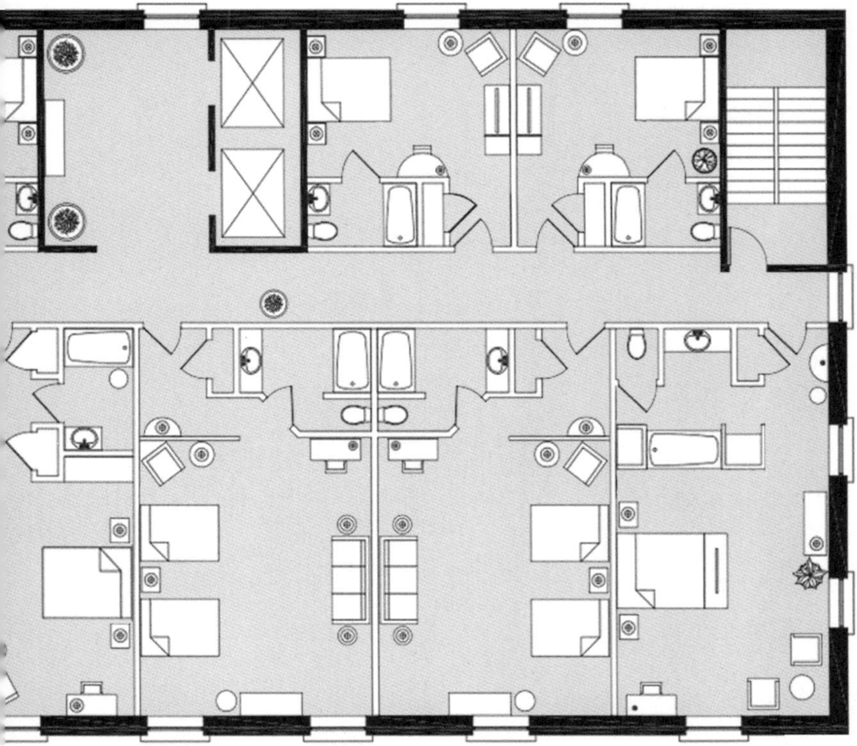

through glass, brick, wood and metal.
- *Find cover from gunfire (e.g., substantial brickwork/ heavy reinforced walls).*
- *Be aware of your exits.*
- *Try not to get trapped.*
- *Be quiet, silence your phone.*

A military medical team performs triage on a casualty. Triage categorizes the priorities for treating multiple casualties.

- *Lock/barricade yourself in.*
- *Move away from the door.*

Tell
- *Call 911—what do the police need to know?*
- *Location—where are the suspects?*
- *Direction—where did you last see the suspects?*
- *Descriptions—identify how many attackers and describe their features, clothing, weapons.*

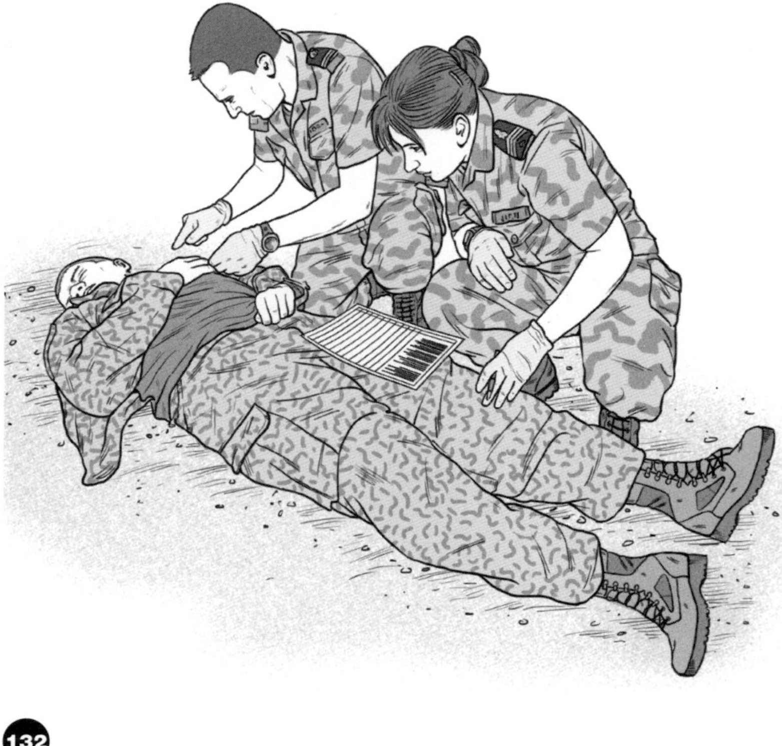

- *Further information—casualties, type of injury, building information, entrances, exits, hostages.*
- *Stop other people entering the building if it is safe to do so.*

When armed response officers arrive on the scene, the government advice is the following:

- *Follow officers' instructions.*
- *Remain calm.*
- *Can you move to a safer area?*
- *Avoid sudden movements that may be considered a threat.*
- *Keep your hands in view.*

Officers may:

- *Point guns at you.*
- *Treat you firmly.*
- *Question you.*
- *Be unable to distinguish you from the attacker.*
- *Evacuate you when it is safe to do so.*

All this advice is essentially sound, although one crucial instruction is missing—fight if necessary. Soldiers and counterterrorist forces recognize that in some instances there is no practical option left for those responding to a terrorist assault but to fight back and bring the attack to an end. For civilians, this is naturally a last resort, but one that will need to be embraced if the passive response is likely to result in your death. There are essentially two main options. The first is to be armed with any weapon you can find—ideally a heavy blunt impact weapon or a stabbing tool—and try to surprise the attacker at close range, deploying the weapon with vicious intent before the attacker can bring his gun to bear on you. If the attack succeeds and you manage to acquire the gunman's firearm (and know how to use it), then you have the opportunity to take out other attackers. As a general rule, however, unless you are a member of the armed forces or security services—and are conspicuously dressed as such—it can be unwise to take up an attacker's firearm, as you are likely to be confused for a terrorist and shot down by the responders.

Although terrorism has become a common worry in many developed societies, we must also fight against the other tool that terrorism thrives upon—fear. Unless you live in the most war-torn of countries, your chances of dying in a terrorist incident are small. One of the greatest victories that terrorists could achieve would be for them to radically modify our daily lives without firing a single shot. From this perspective, "counterterrorism" also includes the mental strength not to be intimidated by terrorism in the first place.

5

Being deprived of liberty is a severe mental challenge. Add additional torments such as ill treatment and even torture, and detention can test the sanity of even the strongest character.

Detention Training

Being captured by an enemy, whether a formal military power or a terrorist group, is a uniquely traumatic event for any soldier, sailor or airman. The captive is plunged into a world of complete powerlessness, with a hostile group of people completely in charge of every aspect of his physical existence. There is an overwhelming sense of dependency—if the captor chose to do so, he could deprive the prisoner of food, water, heat, human company and any form of mental stimulus. Worse still, the captor could potentially torture the prisoner at will (for information or maybe purely for revenge/pleasure) and, as we have seen in numerous recent grotesque cases, have the prisoner executed.

A prisoner has to put in place the mental coping strategies for dealing with imprisonment or from having been taken hostage. I will attempt here to avoid facile advice from someone who has never been held a prisoner. The study of accounts from prisoners and hostages clearly reveals that if conditions are bad enough and torture is a regular feature of imprisonment, then even the most resilient individuals seem to have their breaking point, especially if the incarceration is played out

Opposite: Detention is a uniquely vulnerable situation, the prisoner entirely dependent on the captor.

An overarching challenge of imprisonment is boredom. With little external stimulus, the prisoner has to find internal ways of generating interest.

over a long period of time. Some of the U.S. military personnel captured during the Vietnam War were held in deplorable conditions for many years before their release or their death from maltreatment. In some cases, particular individuals were held in solitary confinement and did not see another friendly human being for months at a time. Periodically, these prisoners were dragged out for another bout of inhumane torture. Yet throughout these deplorable situations, we also find evidence of human resilience and mental ingenuity, which are the subjects of this chapter.

Mental Resistance

Captivity can inflict a progressive state of mental attrition upon a prisoner. A major cause of this wearing effect is actually physical, not psychological, depending on the humanity of the captors. Conditions commonly experienced by prisoners in their new environment include:

- Inadequate daily nutrition, leading to a general loss of health and vitality, triggering feelings of depression and withdrawal.
- Lack of hygiene, particularly the lack of opportunities to wash, shower and change clothing, leading to poor skin and hair condition and infestations by vermin. Poor hygiene also leaves the individual more vulnerable to disease and general ill health.
- Punishing daily routines, including the use of prisoners for hard labor, the exertion of which can be compounded by the poor diet and the lack of general healthcare.
- Periodic violent encounters with guards or other officials, ranging from occasional blows through to systematic and inventive torture. Any injuries sustained are likely to go untreated.
- Extremes of heat and cold. The prison environment is rarely adapted for the prisoners' comfort, hence any cells or barracks might be unbearably hot or cold depending on the season. This problem can be exacerbated by the fact that prisoners are likely to be wearing the clothing in which they were captured (although sometimes they will have been deprived of this), which would only be suitable for the time of year in which they were taken prisoner.

A hooded prisoner crouches in an agonizing stress position. Such positions are used to generate feelings of helplessness and dependence on the captor.

As we can see, the physical conditions of imprisonment alone are sufficient to break mental resistance. But to these we also add the psychological maladies of life in detention—isolation, depression, desperate homesickness, fear, lack of anything to do and uncertainty about the future. Studies into the effects of detention on prisoners of the communist Chinese during the Korean War were conducted in the United States in 1950s and distilled into the publication *Prisoner of War Resistance* (FM 21-78). In the following passage from the publication, the breakdown of mental processes is described but the passage also ends with some suggestions for psychological resistance. Note that the abbreviation "PW" stands for "prisoner of war":

Prisoners often have a definite propaganda value for the captors. On some occasions this value can work to the advantage of the prisoner.

> *Complex mental functions are lost first: the ability to carry out highly creative activities; to meet new, challenging and complex situations; to deal with trying interpersonal relations; and to cope with repeated frustration. These functions can be impaired by a relatively small amount or degree of the factors that upset the body balance—pain, fatigue, lack of sleep, anxiety.*
>
> *As the impairment of brain functions continues, somewhat less complex activities deteriorate. Speed and efficiency in carrying out normal tasks decrease. There is less concerned about "morality" and "right and wrong."*

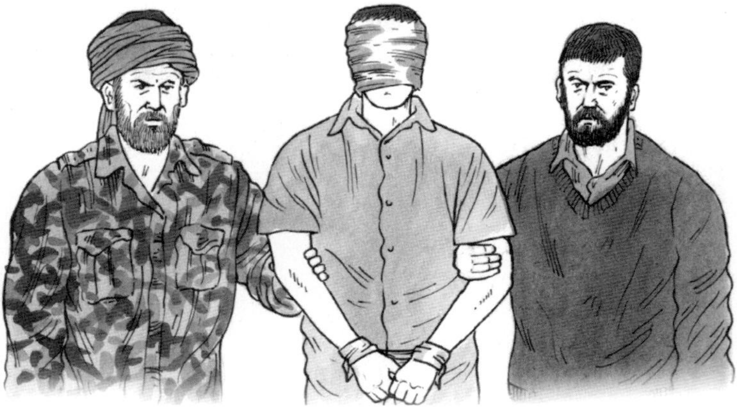

The prisoner is most vulnerable during the first moments of capture, especially if surrounded by a hostile population.

> *Generally, socially oriented behavior falls away. Concern about personal behavior, speech and dress disappears. Emotion takes the place of reason. The needs of the body—i.e., sleep, rest, comfort, desire for freedom from harassment, food and easing of pain—become most important to the individual PW. To some degree, the PW becomes irrational. This is unfortunate, particularly among leaders. Those suffering loss of insight and judgment are totally unaware they suffer these impairments. Thus, they continue to exert their authority, perhaps becoming even more authoritarian and, unknowingly, less sound in their judgments, in effect more irrational. To overcome this problem, all leaders should use their staffs in formulating decisions, particularly those relative to PW policy, conduct, condemnation and punishment.*
>
> *PWs can prevent mental deterioration by playing mental games, working problems in the mind, writing a book—generally keeping their minds active. By so doing, they tend to defeat the captor's system. The individual who allows his mind to stagnate is the one who has problems.*
>
> —U.S. Army, FM 21-78, *Prisoner of War Resistance*, p.17

Meditation is time well spent in detention, as it gives an opportunity for the prisoner to recuperate mentally and physically.

People who have been imprisoned for long periods of time note how prisoners can revert to an almost animal-like state, their horizons defined by little more than the desperate need to acquire food and survive physical punishment. In such situations, prisoner solidarity can all too easily break down, and callousness toward human suffering can dominate. But as the last sentences of this passage indicate, there are some strategies that the prisoner can apply in an attempt to survive captivity, at least mentally. From prisoner accounts, the following seem to be the most common:

Active Social Life

Human beings are incredibly inventive, and historically prisoners have found numerous ways of keeping themselves entertained and stimulated. If they are in communal barracks, and if conditions allow, they can establish discussion groups or tongue-in-cheek societies, stage theatrical performances for each other, make models from the materials around them, write books, sketch pictures, play cards or engage with all possible social activities. The key is to establish regular events or mentally stimulating activities that give structure to the days and weeks, providing a sense of future movement to time. Group activities are particularly beneficial because they give the individual a responsibility within a social context, and so provide a sense of belonging. Team sporting activities are ideal, although opportunities for sport tend to be very limited in modern conditions of captivity.

Active Mental Life

Prisoners held singly or kept in isolation from other prisoners with almost nothing of purpose to do have to fall back on other resources.

Many prisoners in these conditions have testified to the development of a vivid mental life. Yet rather than just indulging in imagination as an escape from their reality, what some did was create a particularly structured mental life with its own internal set of objectives and purposes.

For example, one U.S. airman in the Vietnam War was held for many years in solitary confinement in North Vietnam. During some of the spells of isolation, the officer mentally constructed the house of his dreams, doing so in real time. So if digging the foundations of the house would

By imagining that a wall is a window, through which a beautiful view is seen, the prisoner can mentally "enlarge" his prison cell and make it feel less confining.

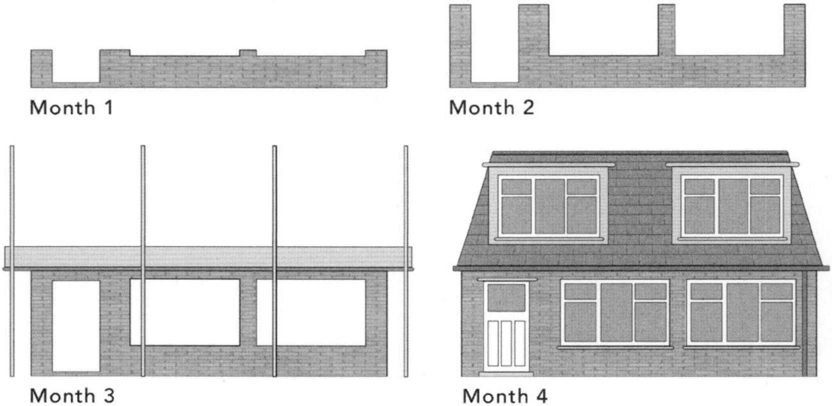

In this mental project, the prisoner mentally builds a house in real time, absorbing several months of his confinement.

take one week in real-time, he would spend one week mentally organizing and implementing that process, paying attention to all the details, such as purchasing the necessary equipment, assembling and instructing the laborers and surveying and mapping out the area. In this way, he was able to absorb many months of his captivity in purposeful mental action.

Building a house is just one example of mental coping strategies. Others could be internally writing a book, mentally painting a picture, setting up and running a business—the possibilities are only limited by the imagination. The key to making such mental activity sustainable is to give it structure and discipline. Yet it is also important not to make the retreat into the imaginary world a complete flight from the real world—the prisoner must always retain an alert connection to his surroundings. At the same time as the prisoner can create an imagined reality, he must also take whatever glimmers of comfort he can from the world around him, however harsh that world might be. The kindness of a fellow prisoner, the sound of birdsong outside the wire or building, a crack of sunlight breaking through an aperture—if the prisoner can draw some sort of positive reinforcement from such small events, then his chances of mental survival increase commensurately.

Physical Activity

As much as he can within the limits of his prison diet and his physical environment, a prisoner should also engage in physical exercise. Not only will this increase his general sense of well-being through the release of endorphins, but it will also maintain the physical fitness required to perform an escape should the opportunity arise. Even in the confines of a relatively small prison cell, basic physical exercise can be undertaken. By doing so, the prisoner might also feel a sense of resistance against the enemy, which can keep him strong even as the enemy attempts to render him weak.

Keeping fit is an excellent morale booster in detention, although the exercise routine needs to be restricted if the prisoner is subjected to food restrictions.

Interrogation and Torture

Interrogation is the attempt to coerce information from an individual by means of duplicity or force. In popular thinking, we tend to associate interrogation with a prisoner being bombarded with incessant questions, the questioning occasionally punctuated by physical brutality

that provides extra incentive for the prisoner to talk. This is certainly one type of interrogation that is commonly used. Long-term questioning, especially when combined with sleep deprivation, can lead to the eventual release of confidential information, more through the weakened self-discipline of a fatigued mind than any lapse in moral fiber.

Yet interrogation can be subtle and diverse. It can include gentle and apparently cooperative conversations between the interrogator and the prisoner, in which the interrogator tries to build up a confessional atmosphere of trust to make the prisoner feel safe and rational in releasing information. Sometimes the interrogation might not appear like an interrogation at all.

For example, the prisoner might casually chat with one of the prison guards before being taken to the interrogation room, not realizing that the guard is consciously probing for information. Conversely, the interrogation may be brutal in the

Sensory deprivation is a cruel punishment, resulting in a loss of sense of time, feelings of panic and depression. One defense is to use a calm and rational inner voice to speak reassuring words, such as "I can cope with this, and I can control my thoughts and feelings."

extreme, essentially just episodes of torture, and the purpose of the violence may be unclear even to the interrogator.

What is important to recognize is that most forms of interrogation have two basic principles governing them—the interaction between reward and punishment, and the application of deception. The former works on a straightforward basis—a lack of cooperation results in punishment (beatings, torture, isolation, etc.), whereas evidence of cooperation (confession or revealing information) will yield either active

The prisoner should identify which guards are the most aggressive and attempt to avoid them or modify behaviors around them.

Captors might subject the prisoner to degrading physical conditions, in an attempt to induce a sense of worthlessness and compliance.

reward (better accommodation, improved/more food, etc.) or at least the absence of the threatened punishment.

Sometimes these two paths might be embodied physically in either "good cop/bad cop" interrogators, or in different locations—a well-lit, clean and comfortable room representing reward, and a gloomy, filthy and threatening underground chamber being a place of punishment and fear. By using these representations of light and dark, the prisoner will progressively seek out methods to increase time in the presence of one and decrease time in the presence of the other, and unwittingly slide into a state of cooperation by divulging information.

It would be easy to say that in such situations the priority is to resist giving away anything important at all costs—nothing more than the traditional "name, rank, and number." Extreme threat or torture, however, especially when combined with sleep deprivation, might over time lead to total emotional and psychological collapse beyond the individual's control, although there are examples of heroic resistance. So what can an individual do to resist and survive interrogation?

One point to remember is that the prisoner's unit or organization will also be aware of the individual's capture, and will generally assume that the prisoner will be made to talk at some point. Therefore they are likely to modify plans and positions accordingly, although this generally will take some time. The prisoner, meanwhile, will receive no new information while in captivity. His objective, therefore, could be to delay giving out any information until that information will be practically useless to the enemy. Thus he might show a genuine interest in the interrogator's line of questioning, appearing to go along with the discussion (to spare unnecessary and possibly damaging violence), but not giving out anything of value for several days or weeks. This approach, it has to be said, is high risk, as a good interrogator will be sensitive to delaying tactics.

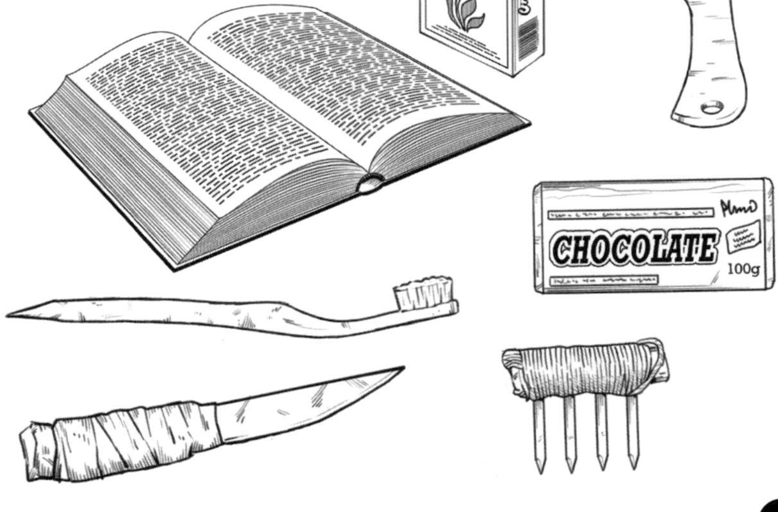

Prison life develops its own "currency," such as books, cigarettes, and chocolate. Everyday objects, however, can also be turned into weapons.

Through staggering the revelation of information, however, leading from the non-sensitive to the sensitive, the clever prisoner might be able to give enough non-consequential but verifiable information to keep the interrogator happy and in the game.

Another possibility is for the prisoner to simply lie. Again, there are risks to this strategy, as the interrogator will have made a career out of spotting people lying. Classic physical signs of someone engaged in lying are:

- Body movements become unnatural, either very fidgety or very still. The feet can become particularly twitchy.
- The liar touches his mouth or other parts of his face/body with his hand, as a nervous gesture.
- In an attempt to convince you of his veracity, the liar will possibly maintain eye contact for a long time without blinking; conversely, those less in control of themselves might look away or up, an instinctive reaction when accessing more imaginative and creative parts of the brain.

A prisoner should be intently aware of body language, his own and others. The man on the left is exhibiting signs of lying or anxiety.

- Speech becomes tighter and less fluid, and breathing becomes more erratic.
- Parts of the face and neck might become flushed and sweaty.

These signs offer good clues that someone might be lying. However, interrogators are now aware that more and more people understand body language. A Special Forces soldier, during their Survival, Evasion, Resistance and Escape (SERE) training, will be coached in how to avoid showing the physical signs of lying. For this reason, a good interrogator will look at more than mere body language to assess the

Mock executions are the captor's demonstration of his total power over the prisoner's life, and have a uniquely traumatic effect on the prisoner's mental state.

veracity of the prisoner's statements. Some of the signals are described in the *Human Intelligence Collector Operations* field manual, and they are worth quoting at length for those who wish to conceal what they know:

- *Internal inconsistencies. Frequently when a source is lying, the HUMINT collector will be able to identify inconsistencies in the timeline, the circumstances*

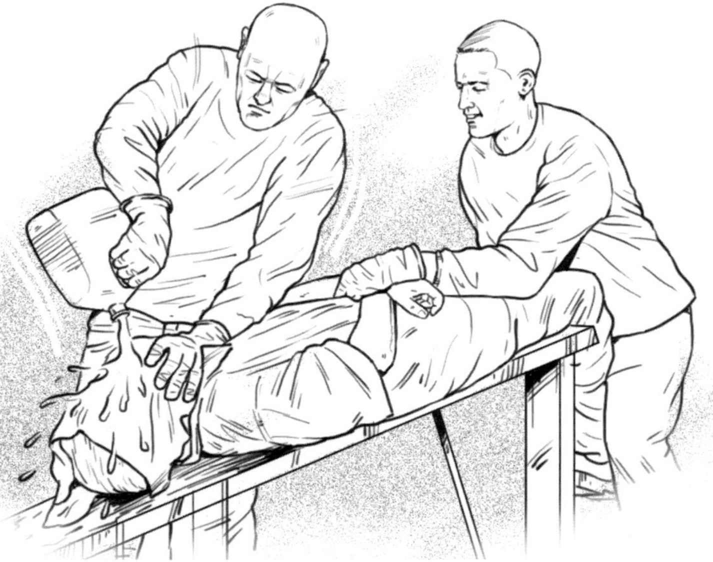

Waterboarding induces the sensations, and the reality, of drowning. The torturer hopes that panic will override the prisoner's restraint with information.

surrounding key events or other areas within the questioning. For example, the source may spend a long time explaining something that took a short time to happen, or a short time telling of an event that took a relatively long time to happen. These internal inconsistencies often indicate deception.

- *Body language does not match verbal message. An extreme example of this would be the source relating a harrowing experience while sitting back in a relaxed position. The HUMINT collector must be careful in using this clue since body language is culturally dependent. Failing to make eye contact in the U.S. is considered a sign of deceit while in some Asian countries it is considered polite.*
- *Knowledge does not match duty position or access. Based on the source's job, duty position or access, the HUMINT collector should have developed a basic idea of the type and degree of information that an individual*

source should know. When the source's answers show that he does not have the expected level of information (too much or too little or different information than expected), this may be an indicator of deceit. The HUMINT collector needs to determine the source of unexpected information....

- *Repeated answers with exact wording and details. Often if a source plans on lying about a topic, he will memorize what he is going to say. If the source always relates an incident using exactly the same wording or answers repeat questions identically (word for word) to the original question, it may be an indicator of deceit. In an extreme case, if the source is interrupted in the middle of a statement on a given topic, he will*

Abusive verbal interrogation is common in detention settings. The prisoner needs to have his own backstory clear in his mind.

have to start at the beginning in order to "get his story straight."

- *Source appearance does not match story. If the source's physical appearance does not match his story, it may be an indication of deceit. Examples of this include the source who says*

he is a farmer but lacks calluses on his hands or the supposed private who has a tailored uniform.

- *Source's language usage does not match story. If the type of language, including sentence structure and vocabulary, does not match the source's story, this may be an indicator of deceit. Examples of this include a farmer using university-level language or a civilian using military slang.*
- *Lack of technical vocabulary. Every occupation has its own jargon and technical vocabulary. If the source does not use the proper technical vocabulary to match his story, this may be an indicator of deceit. The HUMINT collector may require the support of an analyst or technical expert to identify this type of deceit.*

—U.S. Army, FM 2-22.3, *Human Intelligence Collector Operations*, 9: pp.6–7

As we can see, a skilled interrogator will judge all information in a broad and sophisticated context. Not all interrogators are as skilled or well trained, of course, but the prisoner must be careful about making that assumption. One technique to evade detection is to live as close to the truth as possible. By sewing the lie with factual content and

Intelligence officers need to use questioning and observation to separate valuable prisoners from low-ranking combatants.

An abused prisoner needs to control the things he can control: rate of breathing; the content of his inner dialogue; resting when the moment presents itself.

actual memories, it can be easier for the prisoner to tell his story while appearing natural and convincing. At the very least, this approach could deflect more serious physical abuse or divert attention away from other areas of information that the soldier truly cannot divulge.

Ultimately, the prisoner must attempt not to do anything that will put others in danger or lower his sense of self-worth. As the *Prisoner of War Resistance* manual puts it:

> *Using all the traits of character, training, education, pride and faith a man brings with him when captured is his best defense. He must know that captivity is not forever. He must know that he will escape or be released, and that when this happens he will emerge as either more of a man or somewhat less of a man than he was before capture. He must know that he will have to live with this knowledge, and that he will forever be judged by those who knew and know him. He must realize he will always carry his own mirror... his conscience.*
>
> —U.S. Army, FM 21-78, *Prisoner of War Resistance*, p.29

Forming Bonds

As the case of the Vietnam prisoners indicates, imprisonment can be a long-term experience, lasting many months, years or even decades. During this time, it is likely that certain degrees of relationship will evolve between prisoners and their captors. In some cases these relationships can burgeon into the status of genuine friendships, as each party gets to know the other on an individual level. The FBI estimates that about 8 percent of victims of hostage situations actually evidence what is known as Stockholm Syndrome, in which the hostages begin to show signs of empathy and protection toward the hostage takers, even sometimes to the point of defending them physically or verbally.

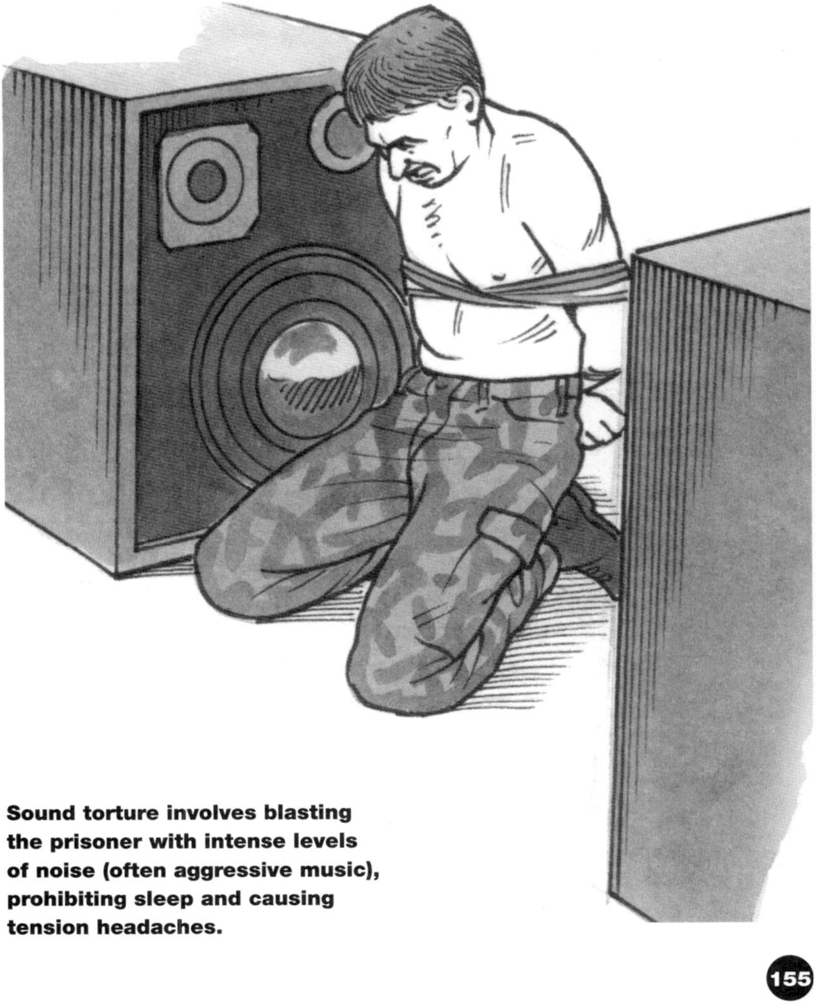

Sound torture involves blasting the prisoner with intense levels of noise (often aggressive music), prohibiting sleep and causing tension headaches.

On several obvious levels, this is a problem. Becoming personally acquainted with the captor can make the prisoner more susceptible to releasing confidential information, betraying other inmates or not looking for opportunities to escape or to assist during a hostage-rescue mission. Yet there are also undeniable advantages. Having an empathetic relationship with captors may lead to better treatment and accommodation, and also the opportunity to acquire information that might be useful to subsequent escape.

In such situations the prisoner must make a clear distinction between the captor as a human being and the position or organization that the captor represents. He should understand that while the captor might display kind qualities and a backstory that inspires sympathy or at least understanding, the fact remains that he represents an organization fundamentally opposed to the prisoner's values and community. By making this distinction, the prisoner will be able to keep a sense of self-awareness about the relationship between them, and possibly be prepared to put aside all sympathy for his newfound friend should a rescue mission emerge.

Yet there also remains the possibility that the captor could be converted to a definite ally if his bond to his organization's objectives is weak. Such opportunities must be explored with caution, because a would-be ally could suddenly switch to a dire enemy if he felt that you were putting him or his family in direct danger.

Escape

The ultimate goal of most prisoners is escape. Successful escapes are, taken in relation to the volume of prisoners, actually quite rare. Not only will the captors use all their intelligence to try to prevent means of escape, but prison life tends to weaken prisoners' bodies and leaves them less capable of making a physically demanding escape attempt.

This does not mean that prisoners should not look for any and all opportunities to escape. The actual techniques of the escape are provided in other volumes (see, for example, *Prisoner of War Escape and Evasion*, Amber Books, 2012). On a psychological level, however, escape is essentially about awareness and information gathering. From the moment of capture, the prisoner needs to be ultra-observant about every aspect of his surroundings, including:

- What do the captors typically wear?
- What is the hierarchy of authority?
- What are the daily routines of the place in which the prisoner is held?

- What vehicles are used to transport prisoners?
- What is the layout of the prison camp or detention center? What are its security features? Are there any parts of the camp hidden from observation?

For a prisoner who has just been captured, the best opportunities for escape are often in the first minutes, especially if his captors are distracted by combat or other surrounding events.

Hostage-rescue troops guide a former prisoner to safety. The rescue is actually one of the most dangerous phases of the captivity.

- What type of landscape surrounds the prison camp? Can its location in the country be identified?
- Are there any parts of the prison camp that are in poor repair or that have been newly repaired or refurbished?
- Where can tools, weapons or other supplies be accessed?
- Are there any areas of the camp that can only be accessed through special privileges, and how does the prisoner gain those privileges?
- Are any guards friendlier and more compliant than others?

By answering these questions and gleaning any other information available, the prisoner can build up a detailed and dynamic picture of the physical structure and daily life of the prison camp. Using this information he can identify possible opportunities for escape. While all such opportunities will obviously include a high degree of risk, that risk can be clearly moderated by identifying the following:

- Opportunities to prepare for an escape, such as implementing a physical fitness regimen or obtaining extra food.
- A means of safely communicating escape instructions to all parties involved. Two useful acquisitions for soldiers are basic sign

Prisoners should observe every detail of a detention center, noting its potential security weaknesses and any routines that could be exploited.

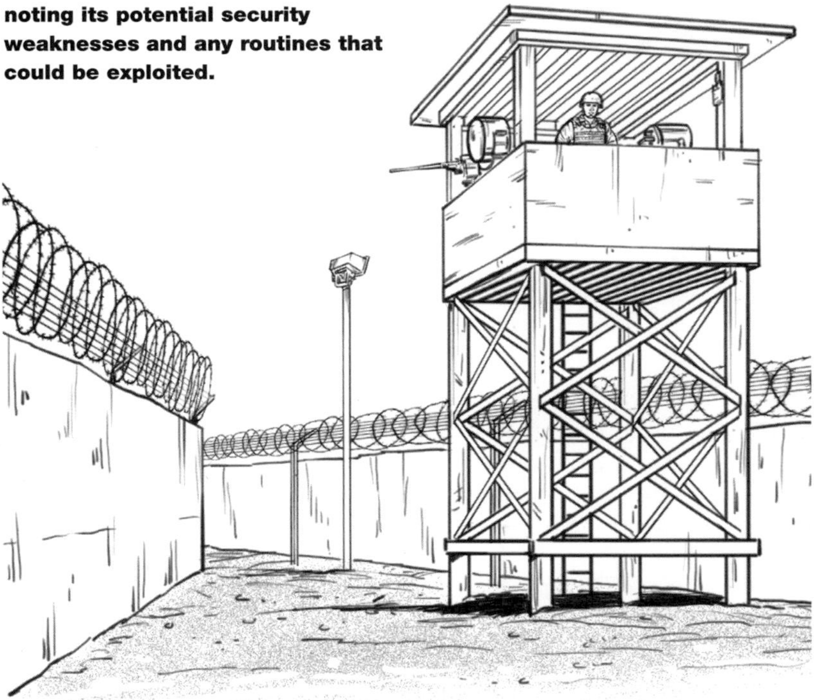

A prison is only part of the escape challenge. The physical surroundings might also be a barrier to escape, especially if they consist of a harsh natural environment.

language for deaf people plus Morse code, which give the opportunities for communication other than speaking.

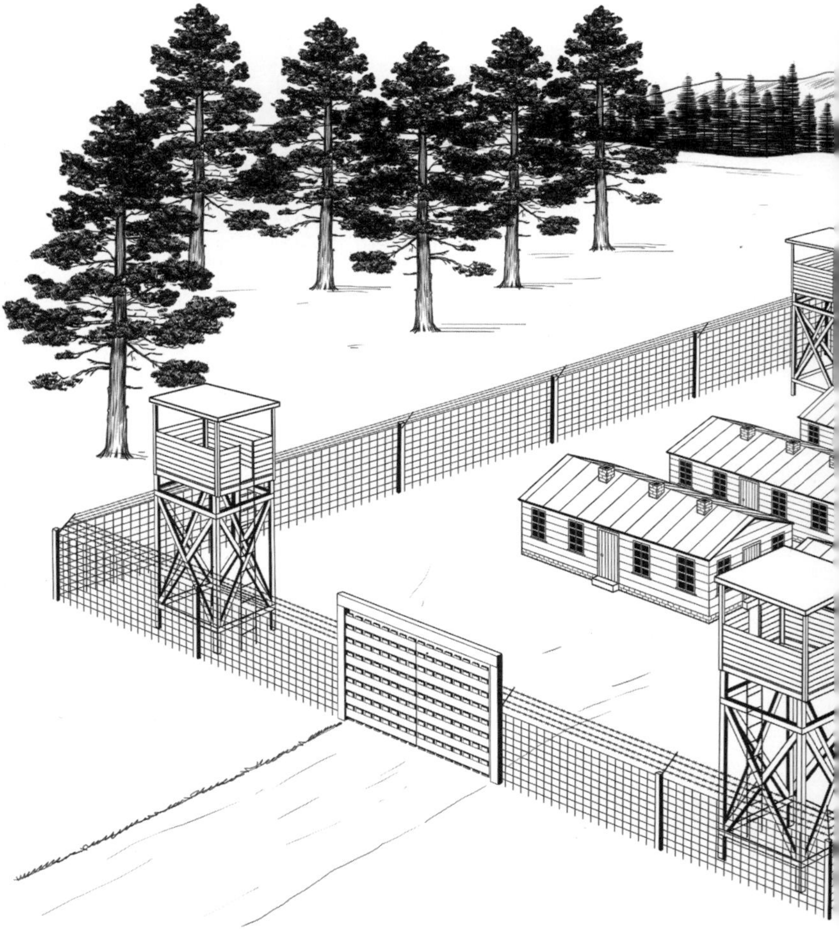

- A physical point of exit from the camp during a window of time in which that point will be unobserved.
- The means by which to make the escape, such as tools or guard uniforms, obtained in advance of the escape attempt and stored until the attempt is launched.

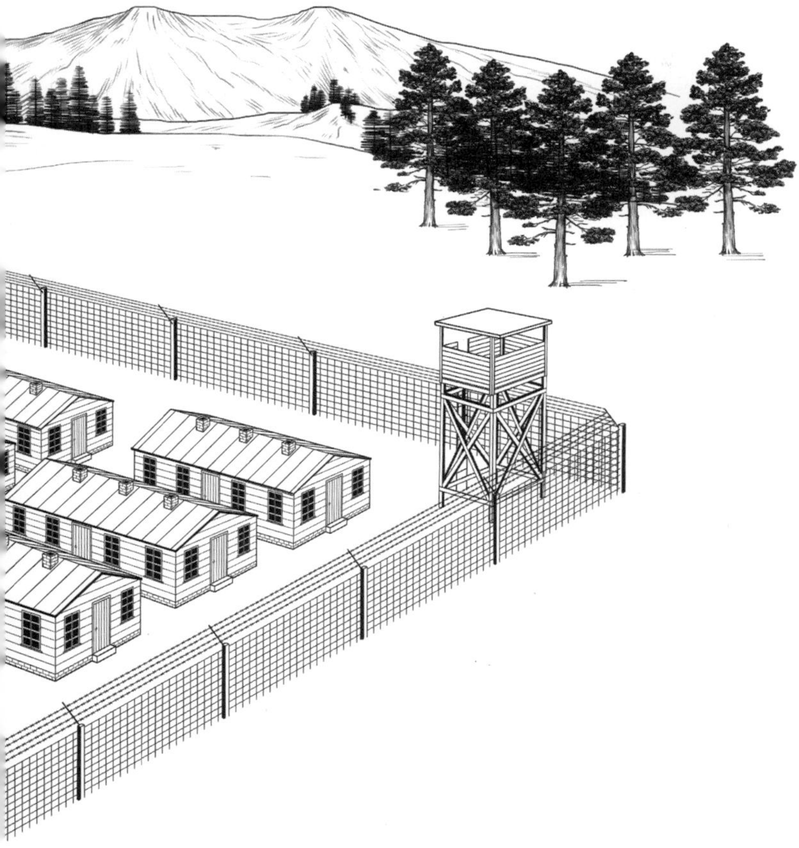

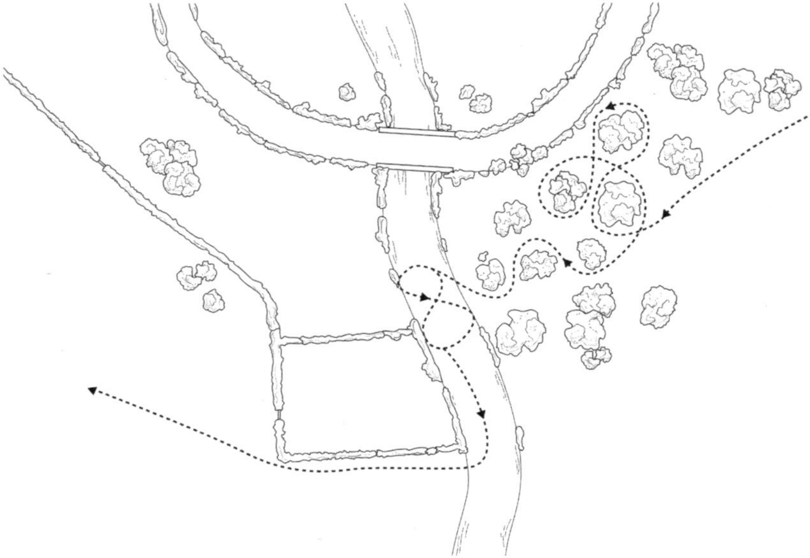

- A clear plan of action to implement if the escape from the camp is successful. In fact, the evasion after the escape can be the most dangerous time, as the prisoner will again be surrounded by potentially hostile territory and any pursuing guards will be more likely to use violence if the prisoner is recaptured. The plan must therefore include a strategy for either contacting rescuing forces or crossing over to friendly lines.

In addition to the planned attempt, the soldier should also stay on the lookout for opportunistic moments

Following an escape, a fugitive can use evasive and irregular patterns of movement to throw off search dog teams to aid his flight.

for escape. These are particularly relevant during the first few hours of captivity, when the captors have typically not yet absorbed the soldier into the detention system and the situation remains fluid.

Once the soldier is in captivity, however, as a general rule planned escape attempts are better because the soldier will have thought through the implications of what he will do beyond the prison wall or wire.

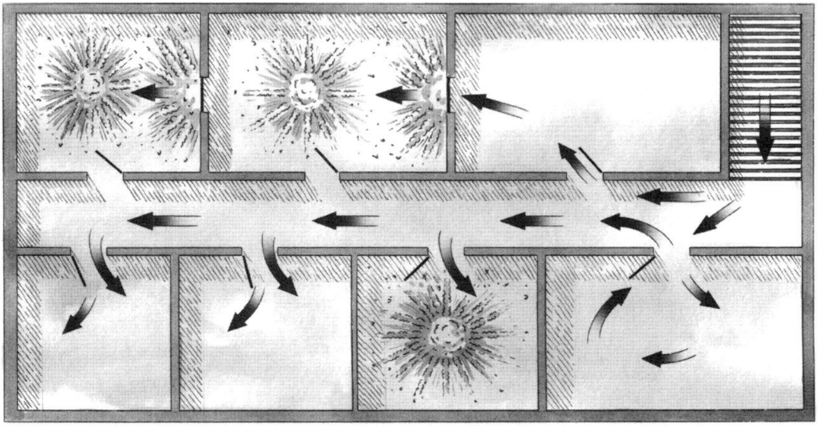

Above: Hostage-rescue teams will move systematically through a building, often using stun grenades to "prepare" the room for entry.

Below: When the rescue team enters a room, the members quickly look for targets, so do not do anything that could be misconstrued as hostile action.

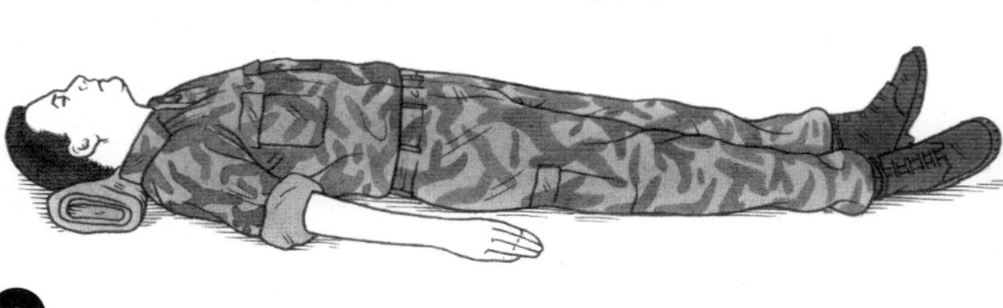

6

After the Battle

The experience of combat has a powerful effect on the human mind, one that can shape the brain at a biological level. The prevention and treatment of trauma is now a top military priority.

One of the biggest areas of financial and intellectual investment in military psychology is the study of what the experience of combat does to the human mind. Although the mental impact of warfare was implicitly recognized as far back as ancient and medieval times, it was during World War I (1914–18) that the study of combat-related psychological trauma became more formalized, as hundreds of thousands of soldiers succumbed to what was labeled by the British as "shell shock." Over time the diagnosis and the labeling were refined, and today we have a very detailed picture of what constitutes "post-traumatic stress disorder" (PTSD). In this chapter, we will explore what happens to the human mind when exposed to the astonishing sights, sounds, smells and sensations of warfare, and also the latest measures used to prevent and treat PTSD.

Hard on the Mind

Not every soldier who goes into combat subsequently suffers from disabling mental conditions. Nor does every soldier react in the same way to the same stimulus. The soldiers of a small-fire team might go through exactly the same engagement, but

Opposite: Combat stress occurs not just during action, but also weeks or even months after.

while some of the members emerge unscathed, others might inherit deep psychological instabilities that lead, in time, to dysfunction, substance abuse or even criminality.

Some of the figures relating to PTSD and related issues are sobering. For example, detailed studies of U.S. military veterans returning from Iraq indicate that 10–18 percent of them were likely to have PTSD on their homecoming. Furthermore, up to 25 percent of the soldiers suffered from depression, with high levels of alcohol use.

Caring for animals has been found to have a restorative effect for traumatized veterans, the animal providing love and responsibility.

Combat stress is clearly a problem that needs to be taken seriously. In 2011 alone, 476,514 U.S. veterans were treated for PTSD. In the UK in 2009, 8500 veterans were serving sentences in British prisons, and a further 11,500 were on probation or parole. Many of these individuals were found to be suffering from PTSD or depression, precipitating the criminal behavior.

Combat is an inherently stressful experience, and can present a soldier with sudden and shocking scenes of violence, mutilation and death, but also a range of other mental fractures. The U.S. Army Medical Course manual, *Psychosocial Issues* (MD0549) lists the stressors of combat as follows:

Simple social time is important to psychological recovery from trauma, as it delivers relaxation and an outlet for feelings.

Some of the types of stress in combat are listed below.

a. Physical/psychological strain. This lowers the overload tolerance, but does not produce battle fatigue unless fear and/or internal conflicts are added (for example, sleeping and eating are irregular in quantity, quality and timing; excessive physical exertion).

b. Fear/anxiety of pain, mutilation, death.

c. Fear of failure/disgrace.

d Grief, rage (loss of friends, hatred of enemy, incompetent leadership and so forth).

e. Ethical limits (killing, firing at noncombatants, leaving patients to die and so forth). The soldier may feel guilty of his own rage or acts.

f. Internal conflict (survival vs. mission, loyalty and ideals—leaving wounded friends behind, short timer and so forth).

g. Boredom, restriction, loss of privacy and so forth.

h. Home front worries, disappointment.

—U.S. Army, MD0549, *Psychosocial Issues*, 2-3

Individually, each item on this list could have serious mental impact on its own, but acting in combination—as they so often present themselves in combat—they can lead to profound mental troubles and the full-blown symptoms of PTSD (see feature box). Given that not all soldiers suffer from PTSD, despite going through

PTSD has a variety of "triggers," including nightmarish dreams and flashbacks initiated by simple physical stimuli, such as a particular smell or noise.

the same or similar experiences as those that do, it is worth attempting to define what might make the difference in terms of frontline management. American researchers looked into some of the contextual factors that might increase the chances of a soldier acquiring PTSD. One of the most obvious factors was the soldiers' exposure to combat, and especially whether they saw other people wounded or killed, or felt themselves trapped in a lethally dangerous situation. Those that suffered from physical injury, such as the loss of a limb or a traumatic brain injury, were also susceptible to full-blown PTSD, exacerbated by the physical symptoms of their injuries.

Substance abuse is a key warning sign of PTSD. The sufferer is essentially using the substance to numb the feelings of trauma and anxiety.

The Symptoms of PTSD—U.S. Department of Veterans Affairs

There are four types of PTSD symptoms:

1. Reliving the event (also called re-experiencing symptoms)

Memories of the traumatic event can come back at any time. You may feel the same fear and horror you did when the event took place. For example:

- You may have nightmares.
- You may feel like you are going through the event again. This is called a flashback.
- You may see, hear or smell something that causes you to relive the event. This is called a trigger. News reports, seeing an accident or hearing a car backfire are examples of triggers.

2. Avoiding situations that remind you of the event

You may try to avoid situations or people that trigger memories of the traumatic event. You may even avoid talking or thinking about the event. For example:

- You may avoid crowds because they feel dangerous.
- You may avoid driving if you were in a car accident or if your military convoy was bombed.
- If you were in an earthquake, you may avoid watching movies about earthquakes.
- You may keep very busy or avoid seeking help because it keeps you from having to think or talk about the event.

3. Negative changes in beliefs and feelings

The way you think about yourself and others changes because of the trauma. This symptom has many aspects, including the following:

- You may not have positive or loving feelings toward other people and may stay away from relationships.
- You may forget about parts of the traumatic event or not be able to talk about them.

- You may think the world is completely dangerous, and no one can be trusted.

4. Feeling keyed up (also called hyperarousal)

You may be jittery, or always alert and on the lookout for danger. You might suddenly become angry or irritable. This is known as hyperarousal. For example:

- You may have a hard time sleeping.
- You may have trouble concentrating.
- You may be startled by a loud noise or surprise.
- You might want to have your back to a wall in a restaurant or waiting room.

—Department of Veterans Affairs http://www.ptsd.va.gov/public/PTSD-overview/basics/symptoms_of_ptsd.asp

But there are other factors, apart from the combat experience, which can heighten the risk of PTSD. Soldiers who are deployed for long periods of time are more vulnerable than those who have shorter postings, the length of time in the field equating to a mental "pressure cooker" effect on the soldiers. Interestingly, people of lower rank and with lower levels of education also tend to be more susceptible, as are those people who are single and have family problems. Furthermore, units that suffer from low morale or poor social support also produce more soldiers with PTSD; interestingly, members of the National Guard or Reserves are more exposed than soldiers belonging to the regular armed forces.

What this information tells us is that the soldiers who are most protected against PTSD are those with the highest levels of training, high levels of unit morale, are well integrated into society with good career prospects, are properly supported by family and friends and have controlled and properly staggered periods of deployment to the frontline. In short, the soldiers most protected against PTSD are those who have their wider mental needs met, which in turn makes them more resilient to handling the

In war zones such as Iraq and Afghanistan, soldiers have witnessed at close hand the effects of war on civilians, often with disturbing outcomes.

traumas of combat. Having said this, PTSD is encountered in veterans from all branches of service and every different type of unit.

Servicemen have a shower. Basic hygiene goes a long way to making the soldier feel a sense of pride and control.

"Collaborative Care"

The behavior of a soldier in the immediate aftermath of battle is important for the prevention of embedded PTSD. Studies conducted over the last decade indicate that after combat, intensive psychological analysis is actually not the best method of PTSD prevention. In fact, this can lead to the internalizing of the trauma and the potential for making things worse.

Far more preferable, it appears, is for soldiers to continue to engage in purposeful activity within the context of a supportive unit. Physical exercise is shown to be particularly beneficial, as are the honest and understanding conversations that can take place between men and women who have just been through the traumatic experience together.

In military circles, this approach is known as "collaborative care," and is especially effective because of the natural way it can evolve, making the soldier feel that he is still in touch with normal life. The use of alcohol to

Opposite: Physical exercise has an undoubted restorative effect. It shakes out many of the effects of stress through relaxing muscles and releasing endorphins.

numb the sensations of the trauma is emphatically not recommended.

One of the key understandings to treating PTSD, if it is identified, is that action should be taken quickly as soon as the symptoms are identified. At a unit level, sound principles of treatment for PTSD have been defined in the *Psychosocial Issues* field manual, based on the BICEPS method (see following pages 176–77).

The principles outlined here are focused on providing the traumatized soldier with enough physical rest to recover, but also enough purpose and social support to keep him or her from sinking deeper into the condition. The treatment is also designed to maintain the pride and

Right: The seeds of PTSD embed themselves almost immediately after a traumatic experience, so preventative treatment should begin almost right away.

Immediate Treatment of Combat Stress (BICEPS)

a. Brevity, Immediacy, Centrality, Expectancy, Proximity, Simplicity (BICEPS) is used as an acronym for the principles of treatment for soldiers suffering from combat stress reaction. Brevity, Immediacy, Centrality, Expectancy, Proximity, Simplicity are explained below.

(1) Brevity. Treatment should be brief, lasting no more than three days. For more extensive treatment, ship the casualty to the rear.
(2) Immediacy. Identify the need for care early. You should not wait for a consultant or evacuation to another facility.

(3) Centrality. Treat a casualty in one location separate from the hospital. This is done to maintain military image and to decrease self-identification as a sick person.
(4) Expectancy. Thorough verbal and nonverbal messages from staff, friends and chain of command, the casualty should expect to return to duty after several days. The casualty should understand that no illness or lack of character exists, only normal stress reaction to combat exhaustion—recovery is assured.
(5) Proximity. Provide care as close as possible to casualty's unit of assignment in order to maintain bonding with unit and maintain support from friends.
(6) Simplicity. The goal of treatment is to return the casualty to combat, not to perform psychodynamic therapy.

b. Other. In addition to "BICEPS," the casualty should be provided with the following:

(1) Rest. This may include sedation for one or two nights, if necessary.
(2) Nourishing and appetizing food.
(3) Group support. This includes a chance to talk to others, compare feelings and observe their recovery.
(4) Professional support. This bolsters the individual's defenses; emphasizes unit integrity, acceptance, and reassurance; and appeals to pride and duty.
(5) Military atmosphere. This includes rank, courtesies, field gear, uniform, performing relevant tasks and quarters in a non-hospital environment.

– U.S. Army, MD0549, *Psychosocial Issues*, 2-5

Soldiers can be reluctant to admit to psychological problems and think they should simply "tough it out." Seeking treatment, however, is always the right course.

self-esteem of the soldier, so that he does not feel that he has somehow let his comrades down.

Treatments

More recent research has also identified two new treatment strategies for those who are suffering from PTSD. The first of these is known as Prolonged Exposure Therapy (PET). During this therapy program, veterans are progressively encouraged to remember and verbally describe traumatic incidents in great detail, even recording their narratives and playing them back to themselves. They're also encouraged to face up to situations that act as triggers for the symptoms of PTSD, slowly adjusting to the sensations and dealing with their fears. Through the narrative process in particular, soldiers learn to look at their experiences more objectively and neutrally, and thereby learn to control the emotions associated with them.

Another associated strategy in dealing with long-term PTSD is called Cognitive Processing Therapy (CPT). This program focuses on educating the soldier about what they have been through, learning

Visualization can be used to help the PTSD sufferer cope mentally with the frustrations of daily life after the battle, as well as reducing stress levels.

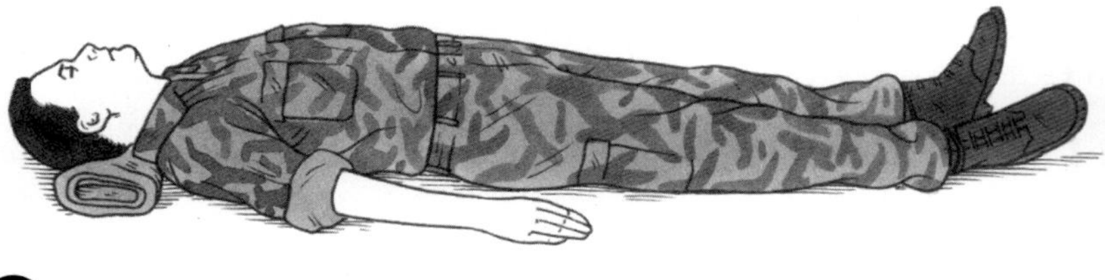

Some of the effects of combat stress can be truly dramatic—here a soldier experiences "hysterical blindness" (the more modern term is "conversion disorder"), in which he temporarily loses his sight.

Experiencing casualties among close friends is a particularly devastating event. The mutual support of the unit is imperative to helping each person cope.

about the symptoms of PTSD and also identifying and understanding the thoughts and feelings associated with the condition. Once they have become more aware of those thoughts and feelings, they progressively challenge them and develop the skills to rise above them. A critical goal of CPT is to stop the PTSD sufferer from looking at the world in terms of danger and threat, and to get them to see positive opportunities for their lives.

A soldier suffering from the symptoms of combat trauma should utilize all the support systems available within his or her unit and in the wider military community. Most soldiers find the greatest source of support comes from their friends and comrades in the service—the people who are typically best placed to understand exactly what the soldier has been through. Yet in the case of psychological problems, the soldier might be unwilling to open up even to his closest friends due to fear of being perceived as unstable and liable to crack under pressure. For this reason NCOs and officers have a particular duty of care to observe their men closely, and to identify those suffering from the symptoms of PTSD or other problems.

Counseling

The soldier might instead turn (or be compelled to go by his superiors) to a professional military counselor or psychologist for evaluation and treatment. This is not something to be feared. Military counselors typically deal with hundreds of cases of combat stress (and other problems) each year, particularly during times of combat deployment, and attach no stigma to PTSD or any other form of mental health problem. It must be remembered that the counselor's objective is to return the patient to active service in a condition of good mental health—the military has invested too much in an individual soldier to want him removed from service.

A good military counselor or psychologist will put in place a program of treatment that aims to help the sufferer understand the source of his trauma, accept and rationalize it, manage his emotional state and move on with daily life. There are also numerous veteran charity organizations that specialize in the treatment of PTSD, many of them

Modern PTSD treatments focus on returning the soldier to normal, albeit non-demanding, duty as quickly as possible, to restore the sense of control and pride.

staffed by war veterans. Indeed, to have full confidence in the treatment given, the soldier is advised to seek out a counselor with actual military experience. The shared background is shown to help generate a more trusting and understanding atmosphere between patient and counselor. It also helps establish a more efficient flow of communication, as the patient and counselor will already possess a shared terminology and less time will be spent getting to understand each other's world.

A word of caution. Not all forms of counseling are of equal value, and much depends on the personal skill of the individual counselor. The soldier undergoing counseling should notice a distinct improvement in his sense of well-being within about a month or so (although severe cases can take much longer to resolve). If not, then the soldier should possibly seek a fresh ear for his problems.

Shared physical activities are excellent for developing a sense of group cohesion after units have been through difficult experiences together.

Life on the Moon

Looking away from PTSD, we also need to address the wider depression that can set in among veterans who have experienced combat, but then later return to civilian life. In many ways, civilian life is particularly hard on soldiers, despite the mental toughness that these individuals will have developed in the services. In the military, soldiers typically experience a great sense of social belonging and common purpose, fostered by a strong unit identity and esprit de corps. During operations, objectives are defined, with each soldier working to protect each member of the team—everyone is watching each other's back. The soldier will be in peak physical condition and will frequently have opportunities to use technologies of enormous power and complexity. Furthermore, all their basic needs will be met within their unit, such as food, accommodation, and employment, unless operational circumstances prove otherwise. In this cooperative context, the soldier

It is recognized that trained psychologists are not always the right people to provide simple counseling. Sometimes guidance comes from other soldiers.

might then go on to experience combat. Although combat can undoubtedly be traumatic, many veterans also remember it for the sheer excitement and adrenaline rush, and for the heightened sensation of being alive—an experience that they never truly capture again.

Set against this background, civilian life can seem unrewarding in comparison. Outside of family, the social bonds between civilians are typically far weaker than the bonds between soldiers, so the veteran can feel adrift in a world with little loyalty and support. Workplace objectives can seem trivial when set against the objectives of a military operation, yet at the same time the demands for career performance in civilian life can be very high and unforgiving if someone does not quite match the company ethos. For those soldiers who have been in combat, civilian life can seem like a terrible anti-climax. In fact, combat veterans have

"Collateral damage" is common even in modern precision warfare. It can make some soldiers question their commitment to the profession of soldiering.

Positive interaction with the civilian population is vital for the success of a military operation, but also for a soldier's mental well-being.

been compared to those astronauts who have been to the moon and subsequently complained of depression, based on the realization that life would never again meet the intensity of the lunar experience.

In terms of treatment for such depression, some veterans have opted for the simplest solution and rejoined the military services, on condition that they are allowed to do so. This approach has its obvious limitations, as at some point the soldier's career will necessarily come to an end, and then he or she will have to face the reality of life outside the military. Far better, therefore, to take some of the principles of military life—principles that the veteran will fully understand—and apply them to the outside world.

For example, operational concepts can be built into civilian goals. The veteran should identify clear "mission objectives" for his life and career—ideally writing them down and pegging them to specific timelines—and initiate a vigorous plan of action that works toward fulfilling those objectives. When the veteran gets out of bed each morning, he should treat each day as part of the mission, making at least some small step toward achieving the larger objective. Ideally, some of those objectives should involve a high degree of social contact, possibly with other military veterans to maintain some semblance of unit cohesion in the outside world. Setting such objectives provides one of the most important things for

Homecoming is a powerful moment for a combat soldier. Studies show that soldiers with stable home lives are less vulnerable to PTSD.

mental survival—a sense of purpose for the future.

Throughout this book, we have seen that military service brings with it a formidable package of mental skills focused on resilience, adaptability, strong body language and getting things done in the real world. These qualities are just as valid outside the armed services as they are in them. In fact, if this book has demonstrated one fact, it is that civilians have much to learn from the mental processes of the military community.

Wounded soldiers have to adjust to new physical realities, such as loss of limbs, in addition to the mental scars that they carry.

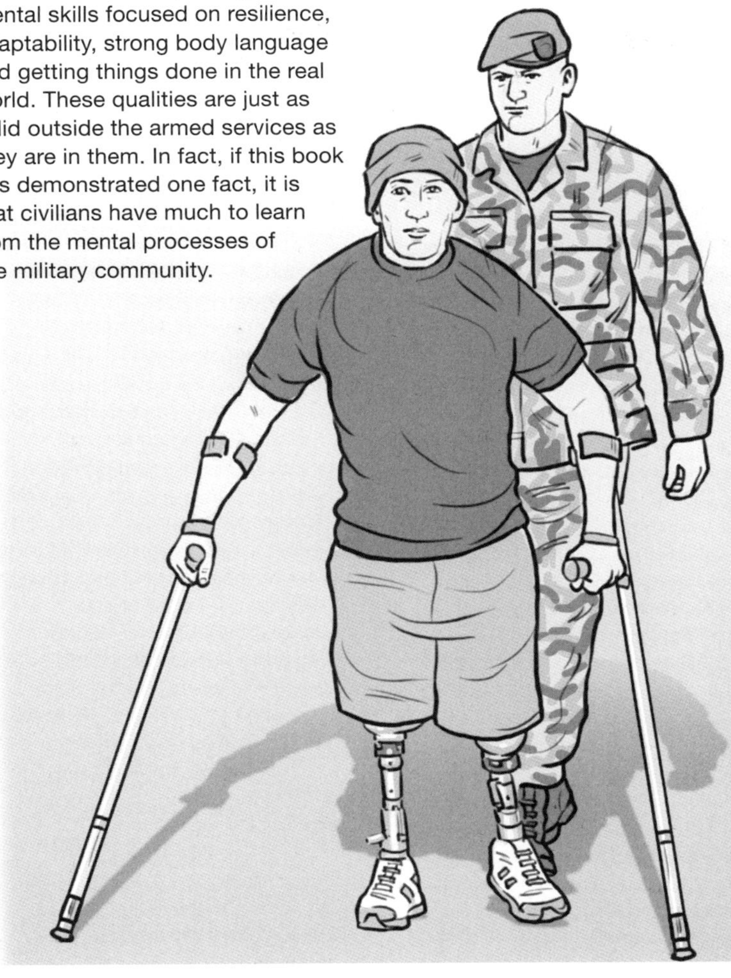

U.S. Field Manuals

U.S. Army, *Prisoner of War Resistance*, FM 21-78 (Department of the Army, Washington DC, 1981)

U.S. Army, *Military Leadership*, FM 22-100 (Department of the Army, Washington DC, 1990)

U.S. Army, *Physical Fitness Training*, FM 21-20 (Department of the Army, Washington DC, 1998)

U.S. Army, *Survival*, FM 3-05.70 (Department of the Army, Washington DC, 2002)

U.S. Army, *Combat and Operational Stress Control*, FM 4-02.51 (Department of the Army, Washington DC, 2006)

U.S. Army, *Human Intelligence Collector Operations*, FM 2-22.3 (Department of the Army, Washington DC, 2006)

U.S. Army, *Insurgencies and Countering Insurgencies*, FM 3-24/MCWP 3-33.5 (Department of the Army, Washington DC, 2014)

U.S. Army, *Leader Development*, FM 6-22 (Department of the Army, Washington DC, 2015)

U.S. Army Medical Department, *Psychosocial Issues*, MD0549 (U.S. Army Medical Department Center and School, Fort Sam Houston, TX)

U.S. Marine Corps, *Kill or Get Killed*, FMFRP 12-80 (Department of the Navy, Washington DC, 1991)

U.S. Marine Corps, *Warfighting*, MCDP 1 (Department of the Navy, Washington DC, 1997)

U.S. Marine Corps, *The Individual's Guide for Understanding and Surviving Terrorism*, MCRP 3-02E (Department of the Navy, Washington DC, 2001)

U.S. War Department, *Bayonet*, FM 23-25, (War Department, Washington DC, 1943)

Books

Clausewitz, Carl von, *On War*, edited and translated by Michael Howard and Peter Paret (Princeton University Press, Princeton, NJ, 1976)

Figley, Charles R. and William P. Nash (eds.), *Combat Stress Injury: Theory, Research, and Management* (Routledge, London, 2006)

Freedman, David H., *Corps Business: The 30 Management Principles of the U.S. Marines* (HarperCollins, New York, 2000)

Laurence, Janice H. and Michael D. Matthews (eds.), *The Oxford Handbook of Military Psychology* (Oxford University Press, Oxford, 2012)

McManners, Hugh, *The Scars of War* (HarperCollins, London, 1994)

McNab, Chris, *Prisoner of War Escape and Evasion* (Amber Books, London, 2012)

Shephard, Ben, *A War of Nerves: Soldiers and Psychiatrists 1914–1994* (Jonathan Cape, London, 2000)

Taylor, Robert L., William E. Rosenbach and Eric B. Rosenbach, *Military Leadership: In Pursuit of Excellence,* sixth edition (Westview Press, Boulder, CO, 2008)

Web References

British Army – Developing Leaders Guide: *www.army.mod.uk/documents/general/rmas_*

ADR002383-developingLeaders.pdf

How to Succeed at Pre-Ranger and Ranger School: *www.benning.army.mil/tenant/wtc/content/PDF/Pre-RangerSuccessGuide.pdf*

Royal Marines – Get Fit to Apply: *www.royalnavy.mod.uk/~/media/files/cnr-pdfs/get%20fit%20to%20apply%20rm.pdf*

UK National Counter Terrorism Security Office: *www.gov.uk/government/organisations/national-counter-terrorism-security-office*

U.S. Department of Homeland Security: *www.dhs.gov*

U.S. Department of Veterans Affairs – PTSD information: *www.ptsd.va.gov/public/PTSD-overview/basics/symptoms_of_ptsd.asp*

U.S. Marine Corps – Semper Fit, Standardized Nutrition Program – *www.dcoe.mil/content/Navigation/Documents/WRC%202012%20Presentations/Family%20Breakout%20Session_King_March%2029_3pm_Meeting%20Room%203.pdf*

U.S. Military Doctrine and Training Publications: *armypubs.army.mil/doctrine/Active_FM.html*

INDEX

References to illustrations and tables/charts/boxes are shown in *italics*